AF412779

Telescope making for beginners

Other books by the same author

Night Skies of the Year (Kahn & Averill)
Exploring the Heavens (Oliver Boyd)
Sky and Stars (Ward Lock)
Exploring Space
The Telescope and Microscope
Looking at Shapes
Time, Calendars and Clocks
Stars and their Legends (Ladybird Books)

Telescope making for beginners

By Roy Worvill
Illustrations by Stephen Weston

KAHN & AVERILL, LONDON

First published by Stanmore Press Ltd in 1974 under
their associated imprint: Kahn & Averill
Copyright © 1974 Roy Worvill
This book may not be reproduced in whole or in part
without permission. Application with regard to any use
of any part of this volume should be addressed to the
publishers.

Revised and reprinted 1976

ISBN 900707 26 7

Printed in England by Biddles Limited,
Guildford, Surrey

Contents

Introduction

In taking up any hobby it is always a mistake to begin with expensive and elaborate apparatus. A much better approach is to start with simple equipment and enlarge or extend it as experience grows. This book is about telescopes and particularly their use for astronomical observation. In this field it is well recognised that the instrument, whatever its quality, is no better than the eye of the observer who uses it. One person will see more with a small telescope than another with a much larger one.

The simple 'spectacle-lens telescope', or small home-made refractor, is probably the best kind to start with. Never under-estimate its powers, for it was this telescope, in the hands of Galileo, which opened the new age of astronomy in the year 1610. It showed the sun's spots, the lunar mountains, the crescent shape of the planet Venus, the satellites of Jupiter, a puzzling glimpse of Saturn's marvellous rings and the crowded star-fields of the Milky Way.

When you have fully explored the possibilities of the small refractor there are many other avenues to follow. There are different telescopes while, for special fields of observation such as comet-hunting, meteor observation and variable stars, a good pair of binoculars is very effective. Photography with the telescope appeals to many amateurs and recently there has been a growing interest in radio astronomy. The scope of astronomy is as wide as the universe itself and just how vast that is, nobody knows.

Roy Worvill
Chipping Norton, Oxon 1973

Glossary of Technical Terms

Achromatic   free from the fringe of colour which surrounds the image formed by a simple lens. In spite of the name no system of lenses is perfect. The mirror, however, is.

Altitude   the angular distance of a celestial object above the observer's horizon.

Azimuth   the angular distance of a celestial body measured between the observer's meridian (north-south line) and the vertical line passing through the object and the zenith.

Altazimuth   a term made by combining altitude and azimuth, meaning a type of telescope mounting which allows free movement vertically and horizontally.

Aluminising   the process of coating a telescope mirror with a thin reflective surface of vaporised aluminium.

Aperture   the diameter of a mirror or lens. The larger the aperture the greater the amount of light collected.

Barlow (lens)   a lens placed in front of a telescope eyepiece which extends the effective focal length of the object-glass and thus increases the magnification of the image with any given eye piece.

Blank   the glass disc for a lens or mirror before it is ground to the required shape.

Catadioptric   a type of telescope which combines the use of a mirror and lens in the formation of the image. The Maksutov is an example.

Celestial Sphere   the apparent dome of the sky to which the stars appear to be fixed and against which the sun, moon and planets seem to move.

Diagonal   the small flat mirror in a Newtonian reflecting telescope.

Equatorial   a form of telescope mounting which has its main axis parallel to that of the earth. When the telescope is directed to the object the observer can follow its motion by one movement of the instrument instead of the two needed by the altazimuth mounting.

Field of View the area of sky visible through the telescope with a given eyepiece. High powers produce a small field of view but there is also variation between different types of eyepiece.

Figure a lens of good figure is one which is accurately shaped to form the best possible image.

Focal Length the distance between the centre of a lens or mirror to the point where light rays are brought to a focus.

Focal Ratio the f/number which is obtained by dividing the focal length of the lens or mirror by its aperture measure in inches or metric units. A lens of 2 inches diameter with a focal length of 20 inches has a focal ratio of 10. It is said to work at f/10. This factor is also used in photography in setting the aperture, or size of the opening, in a camera lens.

Focus the point at which the light-rays from a distant object are brought together after being refracted by a lens or reflected from a mirror.

Magnification the increase in apparent size of an object as compared to that seen by the unaided eye. This is determined by the eyepiece for any given object-glass and is calculated by dividing the focal length of the object-glass by that of the eyepiece: e.g. if the object-glass has a focal length of 40 inches and the eyepiece of 1-inch the resulting magnification will be 40 times or 40x.

Mounting the support for a telescope which enables it to be moved freely and held firmly, an ideal not always easily obtained.

Newtonian the type of reflecting telescope invented by Sir Isaac Newton, still widely used.

Objective the large lens or mirror of a telescope which forms the image. The objective is the vital part of the telescope. A bad one is useless but perfection is rare. Many objectives are in use with minor faults which may be visible only to an expert eye.

Polar Axis the main axis of an equatorial mounting; when correctly adjusted it should point to the celestial pole and it will then be parallel to the axis of the earth.

Secondary Mirror the small mirror of a reflecting telescope.

Speculum an old name for a telescope mirror; the first mirrors were made from an alloy known as speculum metal.

Spider the support for the diagonal (or flat mirror) of a Newtonian reflecting telescope.

Tool   the glass disc which acts as the grinding surface for a telescope mirror. The mirror is placed on top of the tool.

Tube   the means of holding the lenses or mirrors of a telescope may be made from a variety of materials. The refractor usually has a metal or (in simple form) cardboard tube. For the reflector the tube may be of metal, wood or fibre-glass, but is often reduced to skeleton form.

1 The First Telescopes

The maker of the very first telescope is not known with complete certainty. Such evidence as we have, however, points to its invention in Holland, about the year 1608, by a spectacle-maker named Lippershey. Two other men who claimed credit for the new instrument were Zacharias Jansen and Jacob Metius. Perhaps the most surprising thing is that it had not been invented before, since spectacle lenses, with which the first telescope was almost certainly made, had been in use for several centuries and were to be found in several European countries.

News of the telescope seems to have travelled fast and by the year 1609 it had reached Venice. Two pieces of curved glass inserted at the correct distance apart inside a paper tube enabled distant objects to be seen much closer than they appeared to the naked eye. Among the visitors to Venice at the time was a teacher of mathematics named Galileo. He quickly returned to his home at Padua and at once began to make his first telescope.

Galileo looked at the world around him with the eye of a true scientist and saw at once that the telescope might have important uses in extending man's knowledge of the universe. He was too impatient to spend time in making special lenses for his purpose and instead searched the shops of the spectacle-makers to find what he wanted; a good convex lens for the object glass and a concave one (such as the spectacle-makers used for correcting short sight) to serve as the eyepiece. The lenses were placed in a tube made of lead and Galileo's first telescope was ready. It had a magnification of three times, about the same as that of a modern opera-glass, and showed objects at a distance of three miles as large as they appeared to the naked eye at one mile. For his second attempt he succeeded in making one of higher magnification, about eight times, and then he realised that the common spectacle lens was not really good enough for telescope-making. He began to grind his own lenses and experimented with different types. At last he was successful in making a telescope which magnified about thirty times, which is near the limit of magnification for an instrument composed of a convex object-glass and a concave eyepiece. Field-glasses and opera-glasses of this type are still sometimes known as Galilean instruments.

Having completed a telescope of moderate power, Galileo lost no time in putting it to work. 'I betook myself to observations of the heavenly bodies' he recorded, 'and first of all I viewed the moon'. Galileo could scarcely have foreseen what important consequences were to follow his first curious glimpses of the night sky and it is not difficult to imagine with what amazement and wonder he would examine the vast telescopes and other complicated instruments which are now found in the world's great observatories.

The moon revealed its rough landscape of craters and mountains; very like the earth, he thought, but even more rugged. Turning to the planets he saw to his astonishment that they were no longer points of light like the stars. Venus showed its crescent phase, like a tiny moon. Jupiter expanded into a globe and, even more surprising, was the presence of its four large satellites whose movements round the giant planet Galileo followed with growing wonder from night to night. He named them the 'Medicean planets' after the influential Italian Medici family, and re-cognised at once that here was a little model of the solar system as Coper-nicus had described it, with the small planetary worlds revolving around the central sun. The sun revealed its spots, although of course sunspots must have been seen long before the invention of the telescope, for they are often to be seen with the naked eye near sunset or sunrise, or even higher in the sky if the light is screened by dark glass.

The planet Saturn proved something of a problem to Galileo. To be clearly seen, its wonderful ring system requires a somewhat higher mag-nification than Galileo's telescope provided and he was baffled by the changing form of the rings as they present themselves over a period of years.

Galileo's revolutionary discoveries, which contradicted the teaching of the church, aroused the violent anger of the authorities and he was called upon to deny his belief in the Copernican theory about the earth's movement round the sun. He did so, reluctantly, rather than face the dire consequences of a refusal.

One of the disadvantages of the telescope in the form made by Galileo is that the field of view is very small. Only a very tiny part of the sky can be seen in the eyepiece. One of those who saw the need to improve this shortcoming was the famous German scientist Johannes Kepler, born a few years after Galileo. His name is especially remembered for his work in discovering the laws which govern the movement of the planets round the sun, called Kepler's laws of planetary motion. He also made a valuable suggestion for improving the design of the telescope by using a convex lens for the eyepiece instead of the concave form adopted by Galileo. The concave lens certainly had the advantage of showing the object the right way up. This is obviously better if we are looking at people, buildings or the landscape generally, but for use in astronomy it does not matter. When we leave the earth and look out into space we see that our normal concepts of up and down have no real meaning. The fact that Kepler's idea was an important one is confirmed by the fact that telescopes today still use Kepler's arrangement, albeit with certain modifications and improvements. Kepler does not seem to have put his theory into practice, however, and it was another astronomer, Christopher Scheiner, who made the first telescope to the design of Kepler.

This improvement did not mean the end of the telescope-maker's problems. Bigger lenses were being made and these brought new difficulties, particularly two faults which are called spherical and chromatic aberration. The early lenses had curves which were part of a large sphere and light which came from a distant object and passed through the thicker, central part of the lens came to a focus farther from the lens than the rays collected by the outer part. This made it impossible to bring the object sharply into focus. Colour, too, was an annoyance. A single lens cannot bring the light of all the colours in the spectrum to the same focal point. Bright objects were surrounded by coloured fringes of light from the rays which were not correctly focused.

In this respect the lens acts rather like Newton's prism. The blue rays are brought to a focus at a point closer to the lens than are the red rays. Any improvement in the performance of the telescope would clearly have to depend, in the making of the lens, upon overcoming, or at least reducing, the effects of these two serious faults.

One of those who began to search for a solution to the problem was the Dutch astronomer, Huyghens. He began by using a lens of much greater focal length than usual. About 1650 he started work upon his first successful telescope, which was twelve feet long. With this he became the first observer to see the true form of the ring system round the planet Saturn, and he also saw its largest satellite, Titan. This happened in 1656 and these discoveries brought immediate fame to Huygens and his long telescope.

Soon the twelve-foot tube began to seem little more than a toy as telescopes grew to thirty, forty and fifty feet in length. Eventually a lens was made which needed a tube as long as a hundred and fifty feet. It was the work of an astronomer named Hevelius who lived in Danzig. Needless to say a tube of this size presented enormous difficulties and so Hevelius dispensed with it altogether and mounted the lens in the form of what became known as an 'aerial' telescope. Another awkward dilemma was presented by the need to move the telescope in different directions and this necessitated the use of a complicated system of ropes and pulleys.

Clumsy and unwieldy as the 150-feet monster was, it was by no means the largest. Some were made twice as long, but they must have proved almost impossible to use and we have no record of any notable success achieved with them, apart from the discovery of two more satellites of Saturn by the astronomer Cassini at the Paris Observatory. The same observer discovered the main division in the rings of Saturn – this still bears his name – but this was done with a smaller telescope only twenty feet long, and magnifying no more than eighty times. Perhaps the most famous achievement of the long telescopes was the measurement, in 1722, of the diameter of the planet Venus. This was the work of the British astronomer, James Bradley, using a telescope with a length of more than two hundred feet.

The Mirror versus the Lens

It soon became obvious that little more could be achieved by making telescopes ever longer and some other solution was needed if good results were to be obtained with instruments which were of reasonable size. Some opticians and scientists believed that they would be able to make bigger and more efficient lenses if they could find the right materials and the right curved shape for them. Others inclined to the belief that nothing more could be done to improve the object-glass and that the only sensible solution was to design an entirely new kind of telescope. The latter group were much influenced by the opinion of Sir Isaac Newton who had mistakenly concluded that the glass lens could never be freed from the coloured fringes which afflicted the images it produced.

In 1663 a Scottish mathematician named James Gregory had produced a drawing and description of a new kind of telescope which avoided the use of an object-glass. Instead, it formed the image by means of reflecting light from a curved mirror. For many years Gregory's plan remained no more than a theory since no optician of that time could make a mirror with curves of sufficient accuracy.

After looking at Gregory's design Newton decided that such a telescope was possible by using only one curved mirror, instead of the two which Gregory's design required. He completed his first small reflecting telescope in 1668 and, a year or two later, made a second one which he presented to the Royal Society, the body of distinguished scientists to which he was shortly afterwards elected a Fellow.

The telescope, shown in Fig. 11 and still in the possession of the Royal Society, has a focal length of only six inches and was of little practical use for astronomical work. But it was an important step forward in the history of the telescope and the Newtonian reflector has accomplished a great deal in the hands of later observers.

For a long time, however, the reflecting telescope was regarded as little more than a curiosity. Indeed, it took about fifty years to produce a reflector which proved good enough to persuade astronomers to abandon their long refractors. This finally happened when James Hadley made a six-inch Newtonian reflecting telescope. It was six feet long, but, when it was tested, it proved capable of showing all that could be seen in the unwieldy Huyghens refractor which was more than twenty times longer.

Since Hadley's work in the early 1700s, the reflecting telescope has grown in size to the great 200-inch mirror of the Mount Palomar Observatory, and even larger ones in the Soviet Union.

2 The Small Refractor

How it Works

If you have ever used a glass lens to act as a burning-glass then you will
have been using the lens as it works in a telescope. The lens is held so
that it faces the sun. The rays of light are bent, or *refracted*, as they pass
through the lens and will be concentrated into a small bright spot on a
sheet of paper or some other surface held a few inches from the lens.
This spot is actually a very small image of the sun. The large lens of a
telescope works in exactly this way. Behind it the light is collected into
a small image and this image is magnified by another lens which is called
the eyepiece. In this way we get a magnified view of a distant object and
this is what the name 'telescope' actually means – an instrument for
seeing things which are far away.

Optical Parts and Focal Length

In the burning-glass experiment the distance between the lens and the
sun's image, which is focused as a small spot of light, is called the *focal
length* of the lens. This distance is very important when we come to consider the *power or magnification*, of the telescope. The shape of the lens is
also important since it is the curve of the surface and its direction, inwards or outwards, which must make the rays of light bend exactly as
we want them to. The ordinary kind of magnifying-glass curves outwards from the edge on both surfaces and so is a convex lens. Light coming from a distant object, such as the sun or moon, is brought to a point
which is the *focus* of the lens, and at a distance from the lens which is its
focal length. You will see in the illustration how light rays are bent in
different ways by lenses which are shaped differently. Fig 1 (a) shows a
simple convex lens and how it changes the paths of the light rays which
pass through it. The dotted line shows what happens to the blue rays of
light. The two outer lines show that the red rays are bent inwards less

sharply. Fig 1 (b) shows that the concave lens has an opposite effect on the light rays. They are bent outwards and here, also, the blue rays are

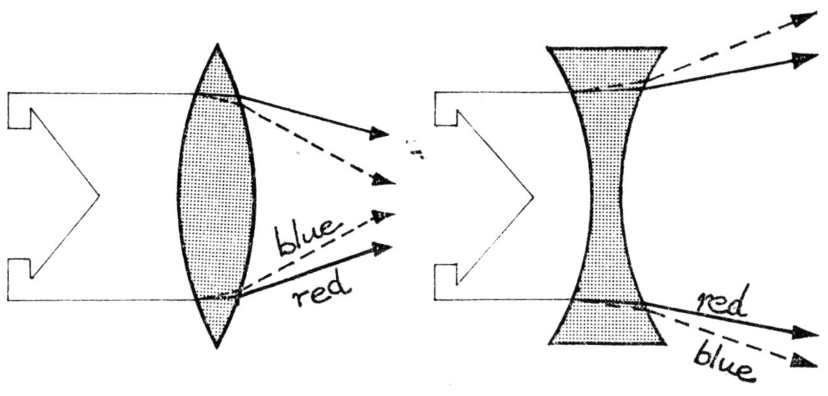

FIG 1 *Separation of colours*

(a) *In convex lens*   (b) *In concave lens*

bent more sharply from their original path as can be seen from the direction of the dotted lines. The change in the path of the light is what we mean by refraction. Another familiar example of refracted light is the appearance of a stick which is partly submerged in water; the stick seems to be bent at the point where it goes into the water. The rainbow in the sky is another example. The rays of sunlight are split into separate bands of colour as they pass through the drops of water in the atmosphere.

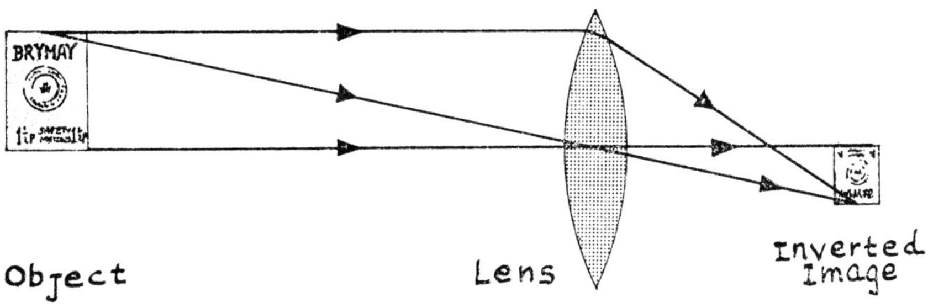

FIG 1 (c) *Image formed by a convex lens*

Since the telescope is used for looking at things which are far away the object we are observing will normally be more distant than the match-box shown in Fig 1 (c). This diagram shows how the convex lens makes an upside-down image of the object as it does when you use a lens as a burning-glass held up to the sun. A camera lens acts in the same way to throw an image on the film and so does the eye.

The Object-glass

The larger lens at the front end of the telescope is usually called the object-glass, or sometimes, the objective. Making a lens of this kind is something which needs a great deal of skill and understanding of the science of optics. Fortunately, although some lenses are very costly, especially in the sizes more than an inch or two in diameter, we can buy convex lenses suitable for our porpose quite cheaply from opticians or from the firms which deal in optical instruments. (Their names and addresses are listed at the back of the book).

Our object-glass should be about two inches in diameter and have a focal length of about thirty or forty inches. The exact figure does not matter. It should be, as far as possible, free from scratches and small chips along the edge are also best avoided, though neither a small edge-chip nor a surface scratch will be visible when the telescope is in use. They merely stop a little bit of light, and one large scratch may be less harmful than a surface which is badly rubbed over a larger area. Incidentally, where optical glass is concerned, it is best to do as little cleaning

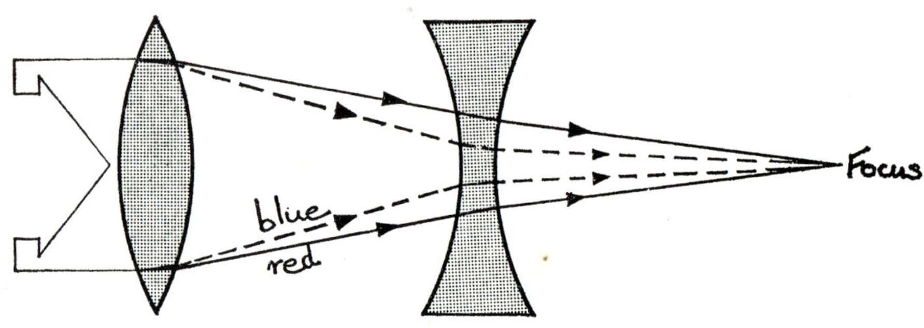

FIG 2(a) *Two lenses can correct colour dispersion*

18

as possible. There are often dust particles present and these do little harm if left alone. Rubbing them off may cause scratches on the lens.

The best kind of lens will be made up of two pieces which may be close to each other or spaced a little way apart. An object-glass of this kind is called achromatic, which means 'free from colour', the colours being those rainbow-like borders which surround the bright objects seen through a single lens. Fig 2 (a) shows how two lenses together do this. The usual form of the achromatic lens is shown in Fig 2 (b). With this lens there will be very little unwanted colour left. The little which does remain will not be serious, especially in lenses of 2 or 3in. diameter.

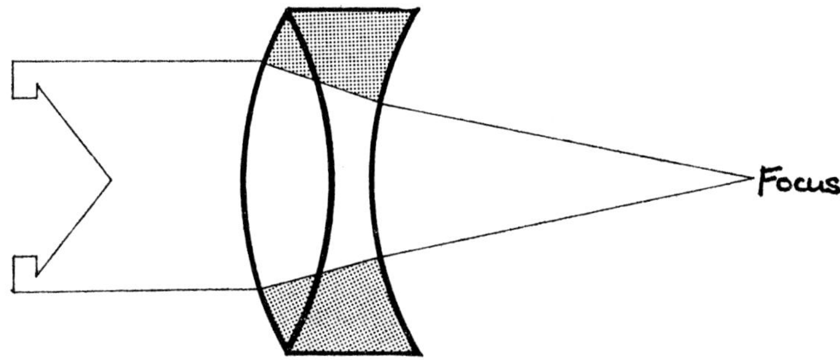

FIG 2(b) *The Achromatic lens corrects colour faults of the simple lens*

The Eyepiece

Our telescope object-glass will produce the image of a distant object but it will be fairly small and so we use a magnifying-glass to enlarge it. This is the purpose of the eyepiece. As you will see in the diagram it is a smaller lens than the object-glass and it also has a much shorter focal length. The lens we shall use for our eyepiece should have a focal length of about one inch.

Magnification

The magnification of a telescope, its 'power' as it is sometimes called, does not depend solely upon either lens. It is decided by the focal lengths of the two lenses forming the object-glass and the eyepiece. If our object-glass has a focal length of thirty inches and the eyepiece lens, (which is another convex lens) has a one-inch focal length, the magnifying power

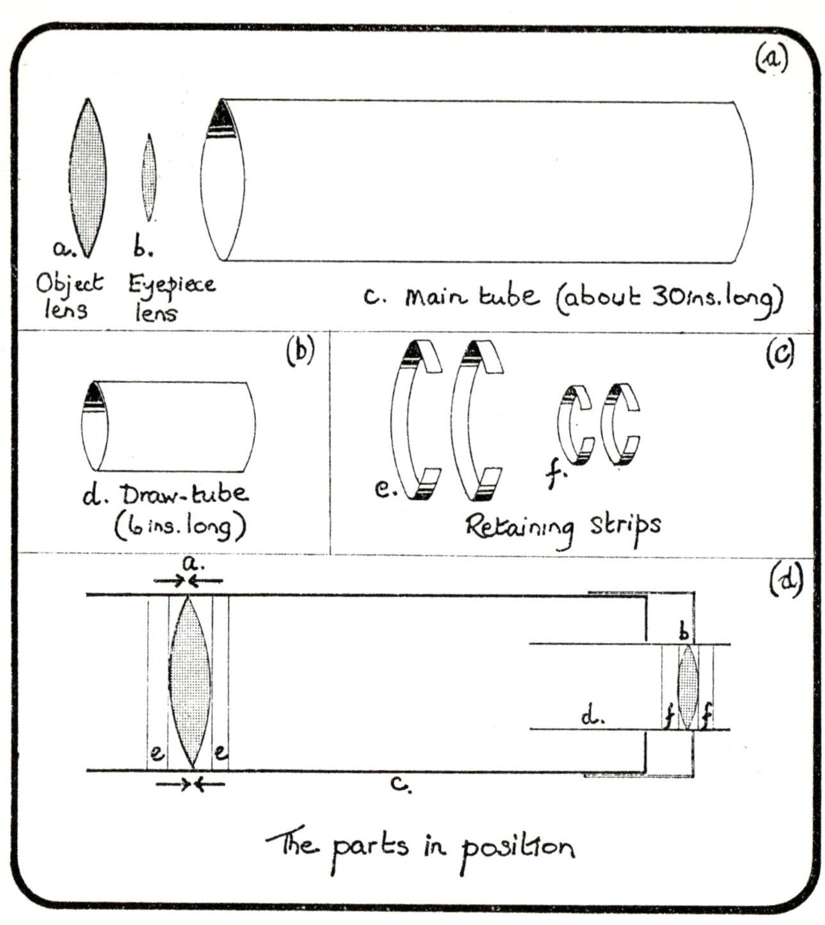

FIG 3

of the telescope will be thirty divided by one, which is thirty times, or 30x, as it is often written. This means that through the telescope we see objects thirty times larger than they could be seen with the naked eye. An object thirty miles away, as seen by the naked eye, will appear then only one mile away. In making this comparison we must remember that

20

if we are looking across the landscape, or across the sea, a great deal of light is kept back by the atmosphere. Thirty miles of atmosphere will stop much more light than one mile.

We can change the power of our telescope by using another eyepiece lens. If this has a focal length of two inches the magnification will now become fifteen (30 divided by 2). The image we see will be only half the size it was with the one-inch eyepiece, but it will be brighter, because the light is now concentrated into a smaller area.

An eyepiece lens with a focal length of a half-inch will give us a magnifying power of sixty. The image will be twice as large now but less bright than before, since we are now spreading the light over a larger area. Perhaps you can now see that there are advantages and disadvantages to both high and low powers. The best course for most purposes is to choose a medium power. High magnifying powers cause frequent disappointment. They may sound impressive but are best avoided for they also magnify the smallest vibration of the telescope and they show a smaller area of sky, called the telescope's field of view. The two lenses we have chosen are about right to begin with. Later on one may enjoy experimenting with higher and lower powers of magnification.

There is one other difficulty which arises from the use of an eyepiece of high magnification – which means one of short focal length. This applies particularly when looking at the night sky. The earth's rotation on its axis from west to east causes the apparent motion of the sun, moon, planets and stars, from east to west. When we use a telescope this movement is very much magnified. The higher the power of the eyepiece the faster everything seems to be moving. A planet or star will very quickly cross the small circular field of view of the high-powered eyepiece. Following it, by turning the telescope with your hand, becomes difficult. This is why large telescopes, which are used in observatories, always have some automatic method of following the stars. Weight-driven clockwork was often used some years ago but more recently an electrical drive has become popular. Even so you will find that a little practice and experience will enable you to keep your object successfully in the field of view without mechanical aids or special 'slow-motion' handles, useful though these are for higher magnification.

Experience will teach you that a 2-inch telescope – which means one having an object-glass of two inches diameter – may be used quite effectively with magnifying powers up to about 60 times, or, as it is called, 60 diameters. But much will depend upon the quality of the object-glass as well as the eyepiece itself. A good achromatic lens of this size would stand a higher power, especially on a night when the air is steady and clear, which is not always the case. Our first telescope with a simple convex lens for its objective would find this a hard test. A very good rule, if you have more than one eyepiece for your telescope, is to begin with the lowest power and then try the higher ones. If you have only one eyepiece make sure it is not one which gives a power of more than about 20 or 30 for each inch of diameter in the object-glass. For a 2-inch this means a power of 40x to 60x. On a favourable night a really good 3-inch object-glass might give a very sharp view with a power of 50 per inch, which would give 150x. Nevertheless, it is as well to remember, that you may see just as much detail on the moon or a planet with a somewhat lower power, say 100x.

It is important to be clear about this matter of magnifying power because beginners often make mistakes through not fully understanding how the telescope works or what the problems are when we think too much about seeing an object as big as possible.

Finally, on the subject of the eyepiece and magnification, you will soon notice that the simple eyepiece made of a single lens, and in fact most eyepieces used in astronomy, give an inverted or 'upsidedown' view compared with what the naked eye sees. So far as viewing the sky is concerned this does not matter. In fact our eye always receives an image which is upside-down but in a remarkable way our brain corrects this by reversing the image. For looking at landscapes or watching ships or animals we obviously want an upright image and the eyepiece needed must produce this by means of two extra lenses or a prism. Either kind of erecting eyepiece can be bought from an optical firm but, like all good optical work, they are fairly expensive.

Assembling the Telescope

You will have gathered that the most important parts of the telescope are the object-glass and the eyepiece. All that remains is to fix these in the right positions and make sure that they can be pointed easily in whatever direction we want to look. The usual way to do this is to put the lenses in a tube of some kind. The first ones were often made of lead. Many different materials have been tried including a variety of metals, such as brass, steel, iron and zinc while wood has also been used a great deal. Some lenses have been mounted without any tube, especially the very long telescopes which were in use three hundred years ago. Some of these had the object-glass fixed at the top of a tall pole round which the observer moved carrying the eyepiece in his hand and trying to get it into focus. In those days astronomers must have had a great deal of patience!

The easiest kind of tube to use for our first experimental telescope is one made of strong cardboard, and good quality postal tubes can be obtained quite cheaply from many stationers' shops. (see Fig 3 (a)). They are normally used for sending charts and other documents rolled up inside them, through the post. Various sizes are obtainable and you will need one with an inside deameter which is a fraction of an inch larger than the diameter of the lens. For a 2-inch lens this means a tube of about $2\frac{1}{4}$ inches for the inside diameter. The length of the tube is the next consideration and it must not be too short. One which is too long can be easily shortened. If the object-glass has a focal length of 30 inches the tube should be about the same length or a little longer.

A smaller sliding tube (see Fig 3 (b)) for the eyepiece lens is required and this needs to be a fraction of an inch bigger in diameter than the eyepiece lens. It will have to slide inside the larger tube but, if it is a bit too small to fit, the gap can be filled in a variety of ways, such as using another short tube of an intermediate size or by having some layers of paper or adhesive tape wrapped round the eyepiece tube. The length of the eyepiece tube should be about six inches. This would allow enough tube to slide in and out for correct focusing.

Mounting the Lenses

It is best to paint the inside of the tubes with a dull black paint of the kind used for blackboards. The purpose of this is to produce a surface which will reduce as much as possible the reflection and scattering of light in the tube. The object-glass should be positioned a few inches inside the tube. The section of the tube in front will then act as a light-shield, rather like the lens-hood which photographers use on their cameras. This avoids stray light coming from the side and, in the case of the telescope, it will also protect the object-glass from any moisture condensing on it, which is often troublesome at night. Even the largest and most expensive object-glass becomes quite useless with a deposit of dew, especially as it is inadvisable to remove it by wiping with a cloth.

A strip of cardboard or other material about $\frac{1}{2}$-inch wide (see Fig 3c) which can be bent to the inside wall of the tube should be glued in position a few inches from the end. The object-glass is then pushed in to rest upon this retaining strip. It is important to place the strip accurately so that the lens will be 'squared-on' as closely as possible to the axis of the tube, since any tilting of the lens will produce distorted images. Another sufficiently thick strip of cardboard is then glued in place above the object-glass to retain it in position with the least possible amount of 'shake'. Two narrow rings cut from the tube will serve this purpose, removing a small section to fit the smaller circumference inside.

Mounting the eyepiece lens is a matter of following much the same procedure, except that the lens must be near the end of the tube where it is easily accessible to the eye (see fig 3 (d)).

This completes the telescope in its simplest form, although there are many possible refinements which will readily occur to the user, quite apart from the provision of some form of mount or tripod. The latter is essential for any optical instrument which magnifies more than about ten or twelve times. Since an object-glass with a focal length of thirty inches and an eyepiece of focal length one inch produces a magnification of thirty times, it is quite impossible to use it effectively without some fairly rigid support.

Apart from the mount, which will be considered next, there are one or two suggestions which may be made. A simple lens, as we have seen, is liable to show some obvious faults, especially in the matter of chromatic aberration, or colour fringes. This can be reduced, and the definition of the lens improved quite noticeably, by inserting a 'stop'. This is made from a ring of blackened paper or cardboard and placed over the lens, leaving a central area clear. The diameter of this hole is a matter for experiment: various sizes can be tried until the best results are obtained.

If, on the other hand, our object-glass is one of the better kind known as the achromatic, composed probably of two elements, and the eyepiece is of similar construction, it may be found unnecessary to 'stop down' the clear aperture. Many optical parts from ex-service equipment can be very useful to the amateur telescope-maker and can often be obtained at much below the cost of new items. Secondhand shops, too, are always worth exploring for optical parts.

3 Mounting the Telescope

A pair of binoculars are easier to hold in a steady position than a telescope. The tube of the telescope is usually much longer and the magnifying power is probably higher. Even a very steady hand is liable to tremble and shake when holding a telescope and the eyepiece will magnify every little movement, so that the object you are looking at seems to be constantly moving about. For astronomical observation there is often a great deal of trouble from the atmosphere and, if the telescope itself is not kept still, any useful observing becomes quite impossible. Any telescope must have a good mount which will allow it to be moved and pointed in any desired direction and also be strong enough to hold it still when we look through it. A vast amount of experiment has been carried out to find the best way to do this and there are many possibilities.

Large telescopes, such as astronomers use in observatories, are almost always mounted on what is known as an equatorial. The main advantage of this is that it allows the observer to follow the movement of a star or planet more easily. This is because the mount is designed to carry the telescope in a curved movement which the object itself follows as it travels across the sky. Amateur astronomers often use this kind of mount too, but for a small telescope there is not much advantage in it. As a general rule it needs to be permanently fixed. The simpler kind of mounting, the altazimuth, is quite good enough for most purposes although we have to make adjustments in both a sideways (azimuth) direction and up or down (altitude).

For a preliminary, experimental mounting to support a light-weight telescope of the postal-tube variety, it is possible to use three strong garden canes, or similar rigid rods, lashed together as shown in fig 4. Avoid four-legged stands of any kind. You will have noticed that many chairs and tables have an uneven stance on the floor of a room. On the ground outside one foot will always be 'dancing' and the smallest movement will result in a shaky image.

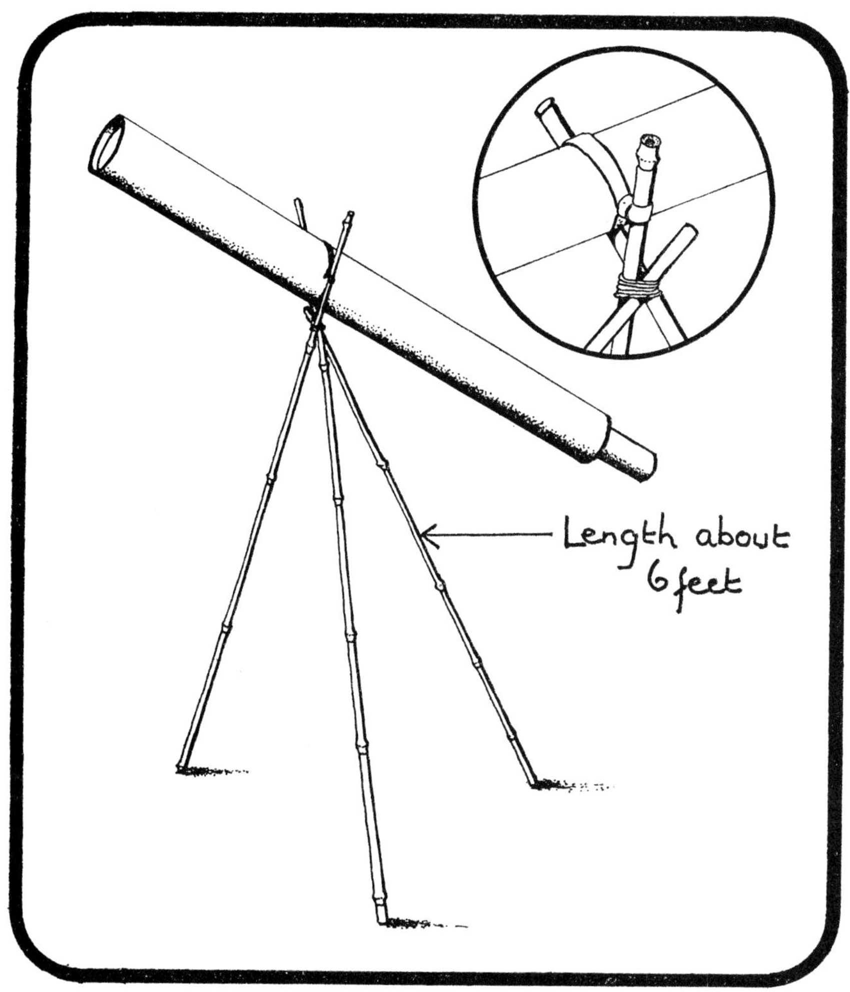

FIG 4 *A simple form of tripod mounting for cardboard-tube refractor*

The short-legged table tripod has also acquired a bad reputation but it is better than nothing and, if fitted with telescopic steadying-rods like some toasting-forks, it can be reasonably efficient. The writer's old 3-

inch Dollond refractor is mounted on a stand of this kind, the pillar-and-claw as it is called, and has proved quite efficient in use. Some observers recommend the beginner to discard such a stand and put the telescope on a tall tripod, but the author of an authoritative work on the telescope, Dr. Louis Bell, records that his own experience of this proved a failure. On some matters even experts may offer quite contradictory advice and the beginner must experiment for himself.

A somewhat more elaborate mount may be contrived by the use of three wooden handles, like those supplied for brooms and garden rakes, attached to a circular or triangular block. A central hole is drilled to take a bolt which secures the upright wooden support to which the telescope tube is attached, either by a short nut and bolt passed through the side of the tube, or by two small straps drawn tightly round a short strip of wood which is hinged by another bolt to the upright.

Home-made mounts of this kind are far from perfect but there is nothing like practical experience for helping the beginner to appreciate not only the advantages of a larger and more elaborate telescope when the opportunity arrives, but also to applaud the persistence and patience of the early telescope users who made their discoveries with the crudest of instruments. As Dr. W. H. Steavenson, one of the most eminent amateur astronomers of our time, has truthfully remarked: "We are all 'primitive methodists' " in the matter of providing our instruments.

Another famous amateur of a generation ago had much the same thing to say about the choice of telescopes for the novice. Arthur Mee, whose book *Observational Astronomy* is still a most valuable guide for the beginner, though now, alas, very hard to find, had this to say after he had acquired a fine 8-inch diameter reflecting telescope by the eminent maker, George Calver: "By this time, however, the charm of novelty had vanished, and, glorious as are the revelations with this telescope, there is no longer the keen delight of the earlier, though infinitely less competent instruments. One seems indeed, to quote Tom Hood, farther off from heaven than when one was a boy. There is a pleasure about home-made apparatus which those who procure their instruments ready-made can never know. And, for the intending amateur, I could wish no other experience than my own".

The small telescope has proved itself in many hands, long since the time of Galileo, Kepler and their contemporaries. To give just one ex-

FIG 5 *Galileo (1564–1642) using one of the first telescopes*

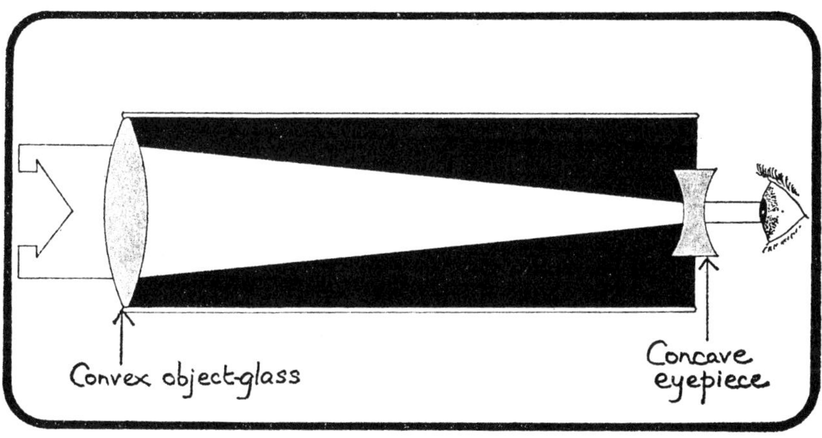

FIG 6 *The form of telescope used by Galileo*

ample, the German astronomer Argelander, who lived a century ago, compiled an atlas and catalogue of over 300,000 stars with nothing larger than a four-inch refractor, and it is a work which is still in use among professional astronomers in all parts of the world. Figs. 5 and 6 show the kind of telescope used by Galileo who was the first astronomer to make use of the optical powers of the glass lens for exploring the sky.

4 Useful Accessories

The mount is an essential part of the telescope but there are some other items which are frequently useful and, as your experience grows, you will be able to appreciate their help. Here are some details of additional telescopic equipment.

The Finder

This is exactly what its name suggests. It is a small telescope with a low magnification and a wide field of view. One can see a larger area of sky with it and will find this very helpful in getting an oject into the eyepiece of the main telescope. It can be made in the same way as the large telescope. One needs a smaller convex lens, about the same size as the eyepiece, but it should have a longer focal length. This means that the surfaces are less steeply curved and light-rays are brought to a focus at a greater distance. A lens of 1-inch diameter and six inches focus would be about right. The eyepiece lens could be of the same diameter or smaller and with a focal length of 1-inch, giving a magnifying power of 6x and the field of view would be quite wide enough to get a star or planet into sight without too much searching. Finder telescopes are rather expensive things to buy and it would not be worth the cost of getting one for your first telescope.

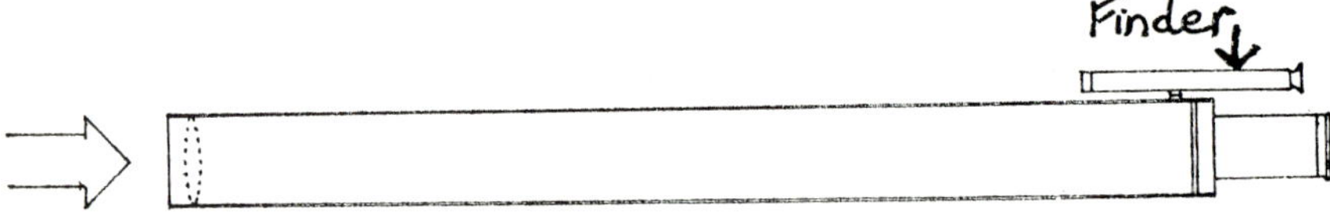

FIG 7 *Finder telescope for refractor*

An even simpler form of finder can be made by fixing a small length of open tube to the main telescope so that it lies parallel to the large tube, like a rifle-sight. (see fig 7) Two small cross wires may be mounted on the eye-end. If the finder tube is made of cardboard, or other material which is not too hard, the cross-wires can be made by pushing two pins through the tube so that they cross in the middle. Even this will be found useful in 'finding' your target in the sky.

The Star Diagonal

Objects in the sky such as the moon, stars and planets are always best seen when they are high up. Some never rise far above the horizon and for these we cannot avoid the troublesome effects of the thick layer of atmosphere we have to look through. If the moon or planet is high in the sky it is much less affected by disturbances in the atmosphere. There is, however, the difficulty of a rather uncomfortable viewing position if we are using a small refractor telescope. It will not take long to become aware of the discomfort of trying to look upwards through an eyepiece which is in a low position.

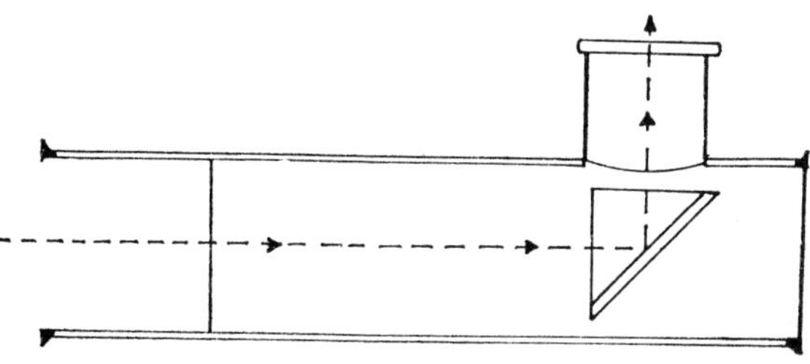

FIG 8 *The Star diagonal*

One way to overcome this is to use the little instrument known as the star diagonal (see fig 8). This consists of a small prism which is mounted in a short, right-angled piece of tube which is pushed or screwed into the telescope draw-tube where the eyepiece is normally inserted. The eyepiece itself is then attached to the star diagonal which allows one to look in a comfortable downward direction instead of having to bend

one's neck which certainly does not help the eyes. There is the slight disadvantage that the prism stops a small amount of light and this is rather
more serious for a small telescope than for a large one which collects
much more light to begin with.

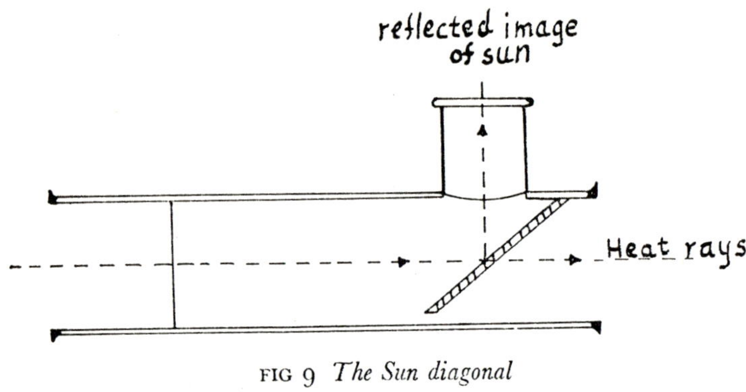

FIG 9 *The Sun diagonal*

The Sun Diagonal

From the outside (see fig 9) this looks like the star diagonal, but inside it
has a wedge of glass which lets most of the sun's light and heat pass
through it and only a small amount is reflected from the surface to form
an image of the sun which is magnified by the eyepiece. As its name
suggests, the sun-diagonal is only used for observations of the sun. But
there are some disadvantages to it. Many astronomical eyepieces were
fitted with screw-on caps of coloured glass for looking at the sun and one
of these must still be used with the sun-diagonal. It is wise never to rely
upon the dark glass cap alone to look at the sun since the heat which is
concentrated at the eyepiece is very great, even in a small telescope.
There is more about observing the sun in a later section of the book.

The Barlow Lens

This is a lens which is mounted in a short tube which slides in the main
telescope draw-tube. It is useful for giving a higher power to any eyepiece with which it is used and the increased power can be varied. The

32

greater the distance between the Barlow and the eyepiece the higher
will be the magnification. In this way the Barlow lens enables one low
power eyepiece to be used for a number of different observations, doing,
in fact, the work of several eyepieces. As with the star-diagonal, however, there is the disadvantage that every lens or prism added to the
telescope stops a little of the light. To get the brightest obtainable image
of a faint object it is best to use as few lenses as possible. Some eyepieces
have been made up of as many as five or six separate parts. One can in
this way get a very wide field of view but, as always, when extra glass
surfaces are encountered by the light rays, the image is made a little
more faint because some light is lost. For a very bright object like the
sun this does not matter but, in the case of a planet, where we want as
much light as possible, the single lens eyepiece may be better for a small
telescope.

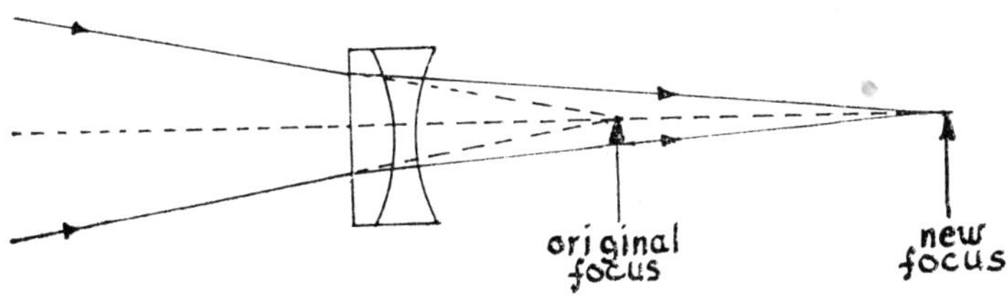

FIG 10 *The Barlow lens*

5 Making a Reflecting Telescope

We have already learnt that a simple lens will not bring all the various coloured light-rays to the same focal point. Some colours, notably the blue rays, are bent more easily than the red ones which have a longer wavelength. When you focus your eyepiece lens on the image formed by one colour, the other rays of light are still there unfocussed and make a blurred, coloured fringe to the object being examined. Some of this trouble can be avoided by covering the edge of the lens or stopping the lens down, as it is called. Still, we cannot completely escape this problem of chromatic aberration (as opticians term it) by any trick with lenses.

Three hundred years ago the great mathematician, Sir Isaac Newton, was much troubled by this difficulty and it was in this connection that he made one of his rare mistakes in thinking that the lens telescope, the refractor, could not be made to work well because of the false colours. He decided, instead, to use a curved mirror to replace the telescope object-glass. Light rays which are reflected back from a curved mirror, like a ball bouncing from a bowl, would make an image of an object at the focus just like the lens did by bending, or refracting the light. Moreover the reflecting surface of the mirror has this important advantage. All the colours are brought to the same focal point. There is no out-of-focus coloured fringe of wasted light. Newton made his small telescope, which is still in the possession of the famous scientific body called the Royal Society, and his design has been successfully used ever since by opticians and amateur astronomers alike. You can see this famous little telescope in fig 11.

The Parts of The Reflector

The Newtonian reflecting telescope, named after its designer, has three main parts. These are

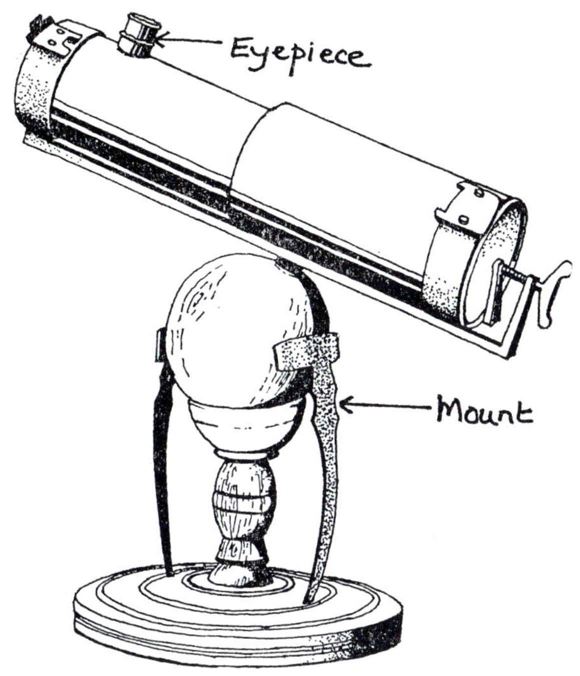

FIG 11 *Isaac Newton's reflecting telescope made in 1671*

1. the main mirror, sometimes called the objective, which collects the light and forms the image.
2. the small flat mirror which bends the reflected light through a right-angle so that it passes out through the eyepiece draw-tube
3. the eyepiece which performs in the reflector exactly the same function as it does in the refractor; i.e. magnifying the image.

These three parts have to be fixed in the correct positions and is usually done, as with the refractor, by mounting them in some form of tube, though it may be no more than a skeleton framework. You will see the path of the light rays through the Newtonian reflector in fig 12.

The Main Mirror

The mirror of a reflector is normally made in larger sizes than the lens of the refractor. One reason for this is that there are some important advantages to be gained by a bigger surface. It will collect more light

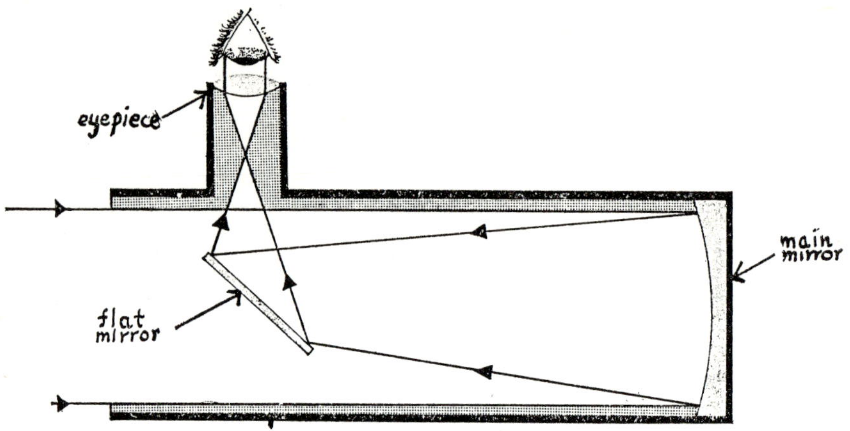

FIG 12 *The light-path in the Newtonian reflector*

and this will give an image which is brighter than that obtained from a small lens or mirror. It will also show more detail in the image. Using a high magnifying power on an image from a very small lens is of very limited value. It is like blowing up a balloon which has a picture painted on it. You may stretch the balloon and get a bigger picture, but the detail is no more nor is the picture any better.

Another reason why mirrors are made in larger sizes than lenses is that the glass of which they are made is much cheaper. Many things, such as port-hole lights from old ships, have been turned into telescope mirrors. A lens has to be made of high-grade optical glass.

The mirror's only function is to give the right shape to a thin film of metal, usually silver or aluminium, to reflect the light. Any faults with the glass, such as bubbles formed when it is made, are unimportant since they do not affect its performance. In the lens they would be more serious because the light must pass right through it.

Making a Mirror

It is possible to buy the optical parts (the main mirror and the flat) of a Newtonian reflector for about £10, if it is no more than six inches in diameter. But there is much satisfaction to be had by making your own and this depends more on care and patience than upon any special skill. The various materials, including the disc of glass for the mirror, can be obtained from many of the suppliers listed at the end of the book.

Reduced to its bare essentials, the job of mirror-making consists of taking the flat disc of glass and grinding one surface into the correct curved shape. This is done by means of another glass disc, the tool, and the grinding effect produced by hard grains of carborundum powder sprinkled between the two surfaces, as the upper disc is moved backwards and forwards across the surface of the lower one. After the mirror disc, which is the upper one, has been hollowed out to the right shape, it is polished by another powder, optical rouge, and then given a bright reflecting surface by depositing on it a layer of silver or aluminium. The process is easily described but it is one which needs a good deal of care. No part of the work should be hurried. Many beginners have made a good mirror at their first attempt and we are very fortunate in having ways of testing each stage to be sure everything is quite right before continuing on to the next stage. The early mirrors were made of speculum metal. This was a mixture of tin with copper or brass and many experiments were conducted to find the best mixture. When the famous astronomer, Sir William Herschel, began making telescopes he had more than a hundred failures before producing a good one. But Herschel lived two centuries ago, before our methods of testing were known.

Rough-grinding the Mirror

Many people have begun by making a six-inch or even an eight-inch mirror, and have then gone on to make larger ones of ten or twelve inches in diameter. It must be remembered that there is a lot more work involved as the size of the mirror increases and, even when it is completed, the problems of mounting the optical parts successfully are bigger and there are problems in using it. Also the atmosphere proves more troublesome with larger telescopes.

A six-inch mirror has some advantages. It is large enough to give a good performance in use without being too cumbersome to handle. But the beginner may find it easier to start with a smaller four-inch size for the following good reason. Returning to our lens telescope for a moment, you will remember that the focal length of the lens was thirty inches and the diameter was two inches. The first figure divided by the second gives us a factor, or ratio, known as the f-number: In this case it is f-15. (A camera lens is marked with a figure like this: f/3·5 or 2·8 for example, and the meaning here is exactly the same). A six-inch mirror of this f-number would have a focal length of 90 inches.

Apart from the two glass discs the other requirements are the following grades and quantities of carborundum powder: Grade 60 or 80, 1 lb. and half of this quantity of the finer grades, 100, or 120, 220, 400 and 600. The precise grades are not important provided we have enough of the coarse grade to do the rough-grinding of the disc to approximately the required curve and then sufficient of the finer grades to remove the deep scratches and pitted surface produced by the larger grains.

The mirror-maker must also have a firm, small table, stool or barrel round which it is possible to move easily while doing the grinding. The other materials are a small quantity of turpentine, about a pound of pitch, a small quantity of beeswax and a few ounces of optical rouge for polishing the mirror. The type of rouge known as cevri is cleaner to handle than the ordinary variety and has the advantage of polishing more quickly. Some of the suppliers listed offer the complete range of materials in kit form.

The first operation is to secure the tool disc to the top of the stool, or other support. This can be done by pouring a little melted pitch on the surface and then pressing the tool disc into it. On a wood surface the tool can be secured against lateral movement during grinding by three or four small blocks of wood nailed or screwed down. Some workers prefer, as a handle, a hemisphere of wood, half of one of the wooden balls thrown at coconuts in a fairground. This can be attached by melted pitch. Others prefer to hold the disc by its edges. This is a matter of choice. Try the latter first and see how you get on.

Only two kinds of surface can be moved across each other with a sliding motion and still maintain a perfect contact while doing so. Flat or plane surfaces will obviously allow this. The other is a spherical surface. Where two such surfaces match (i.e. the inner, concave surface of one subject has the same radius of curvature as the outer, convex surface of the other) they can be moved across each other while still touching at

every point. This is what happens with the grinding operation involving
the tool disc and the mirror. They begin as plane surfaces but during
the operation of grinding, the mirror surface (which is the lower surface
of the top disc) begins to take on a concave, spherical shape and the tool
(the upper surface of which is beneath the mirror) assumes a convex,
spherical form. The action of grinding the mirror is shown in fig 13.

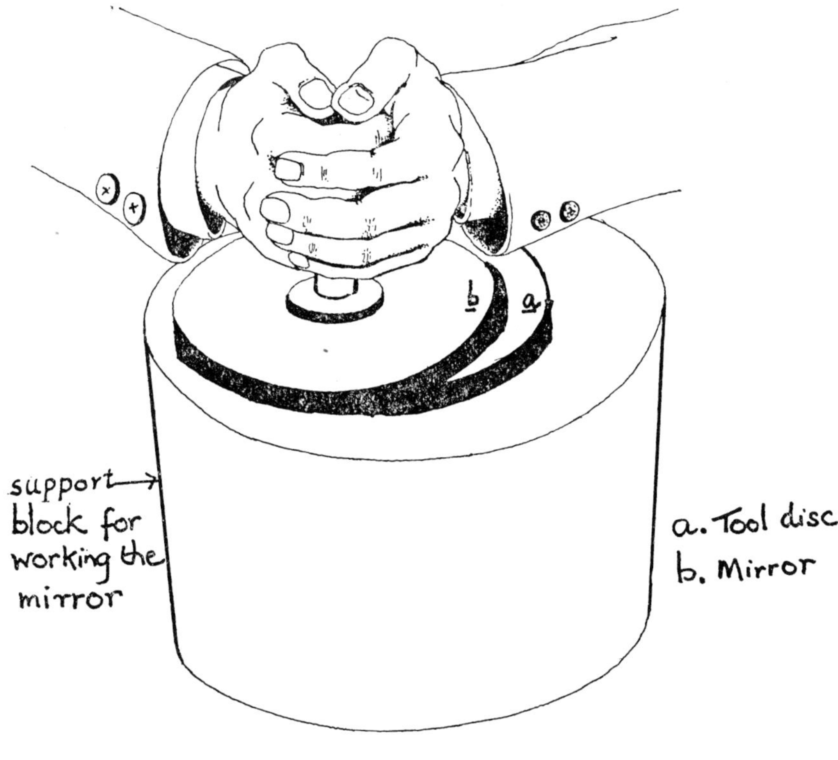

FIG 13 *Grinding the mirror*

The operation commences after a small quantity of the coarse carborundum powder, and then some water, is sprinkled on the tool disc.
Alternatively the mirror disc can be dipped in water. The top disc is
then pushed with an even, sliding movement backwards and forwards
across the tool disc while the worker moves slowly round the circum-

ference so that the direction constantly changes. The mirror, too, is turned slightly after every few strokes. As the mirror disc is pushed backwards and forwards across the tool the 'overhang' at the end of each stroke produces a situation in which the maximum grinding effect is applied at the edge of the tool and at the centre of the mirror. This causes the desired effect of wearing down the edge of the tool and deepening the hollow towards the centre of the mirror. The rotation of the mirror and the movement of the worker tend to spread the grinding action evenly round the mirror's circumference.

The surfaces in contact will soon begin to produce a curved shape and a texture which looks like ground glass. The carborundum powder should be changed whenever you know, by the smooth 'feel' of the strokes, that the cutting action is slowing down.

The Radius of Curvature

After about half an hour or so it is desirable to make a rough test as to how the focal length of the mirror is progressing. With a four-inch mirror, a focal length of about four feet is very convenient. We say that the mirror works at f/12.

To find the focal length of the mirror at any stage in the grinding operation we must arrive first at the figure for the radius of curvature, which is the radius of the sphere of which the mirror disc is part. The focal length will be half of this figure, so we are aiming at a radius of curvature as close as possible to 96 inches. Further fine grinding will have the effect of shortening the focus and so it is as well to allow a margin for this and look for a radius of curvature of about 100 inches.

We find the figure by setting up a small oil-lamp, torch bulb or candle in a dark room and, after wetting the surface of the mirror to give a little more reflecting power, set it on edge, facing the lamp at a distance of about eight feet. Slide the lamp slowly from side to side, a few inches each way in front of the mirror, and notice carefully which way the reflection of the lamp moves. If the reflection moves to the left when the lamp is moved to the left the lamp is inside the radius of curvature. Immediately the lamp is placed outside the radius of curvature the reflected image moves to the right when the lamp moves to the left. By trial and error it can soon be established where the centre of curvature lies and the distance measured and then halved for the focal length. The appearance of the shadows inside, outside and at the radius of curvature is shown in fig. 14.

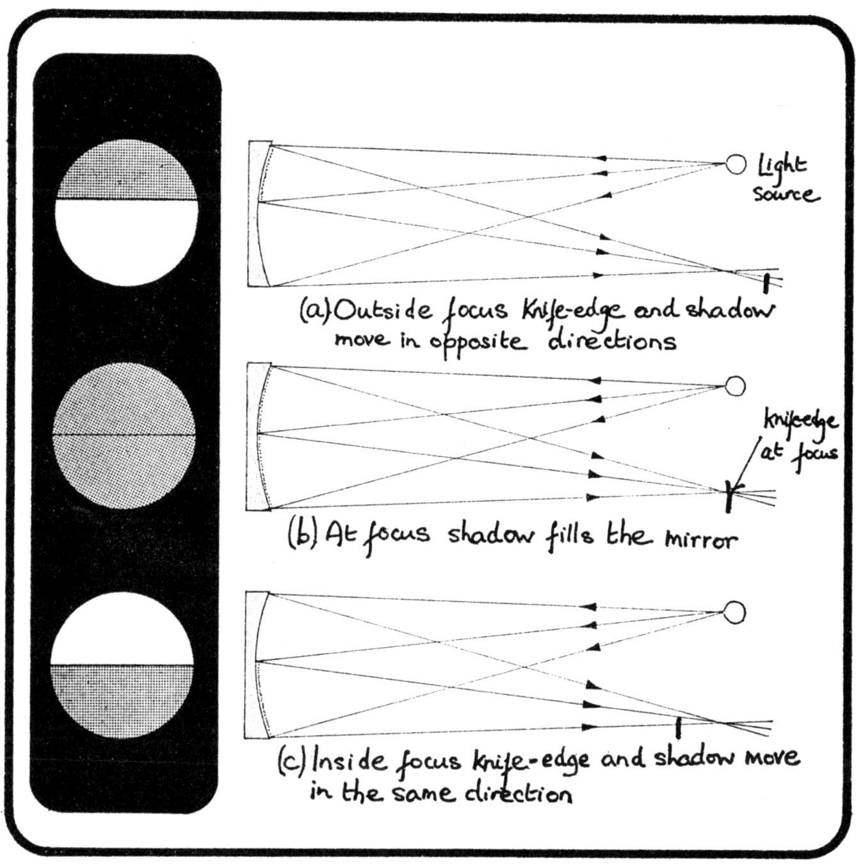

FIG 14 *Shadow appearances in Foucault test for finding radius of curvature of Newtonian mirror*

When the radius of curvature is a little more than 100 inches it is time to remove all traces of the coarse carborundum grit by careful washing under a tap. An old nail brush will help to remove any obstinate particles. Do not be tempted to skimp this part of the proceedings. One or two odd particles of coarse powder lurking in some small pit may cause unwanted scratches in the next stage of fine grinding, which is continued in the same manner as before until the focal length, when tested by the lamp image, is a little more than the required figure. There is, of course, nothing magical about precise whole number focal ratios. It will make little difference whether the final outcome is a mirror of 46 inches or $50\frac{1}{2}$ inches focal length. The total period required to complete the grinding process should not be much more than about four hours. Before embarking on the important job of polishing the mirror surface it is vital that all traces of the carborundum grit be thoroughly washed away and the mirror left clean and dry.

Polishing the Mirror

This part of the work is much like that of grinding but the polish is given to the mirror surface by using the rouge instead of the carborundum powder. First, however, the tool is covered with a layer of melted pitch mixed with a small quantity of beeswax. Then the mirror is pressed down on the tool so that the soft layer of pitch takes the same curve as the mirror. This is allowed to cool before removing the mirror. Then a series of v-shaped cuts are made in the pitch to form a network of squares. The central point of the tool should be marked so that the point comes into a corner of one square but not actually in one of the cuts. If care is not taken about this, the polishing operation may produce a series of rings on the mirror. The action of polishing is shown in fig 15.

In moving the mirror over the tool use somewhat shorter strokes than in the grinding process. In about half-an-hour, or a little more, the mirror should be acquiring a much smoother appearance and it is as well at this stage to check, as before, the radius of curvature. This will be very near the final value and the focal length will be close to the figure arrived at since the remainder of the polishing will not change it by more than a small fraction of an inch. This stage of the operation requires great care in excluding dust or particles of abrasive material from the mirror and keeping, as near as possible, to a constant temperature. Do not, for example, switch on any form of heating while in the middle of polishing.

Polishing may take several hours but this part of the work cannot be hurried without the risk of spoiling the mirror.

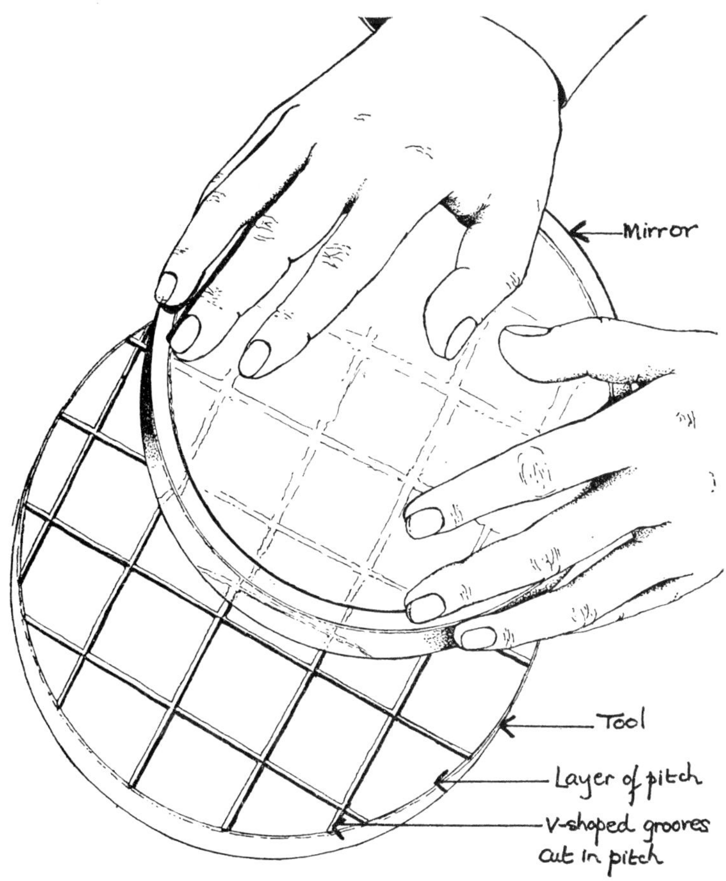

FIG 15 *Polishing the mirror*

The Sphere and the Paraboloid

When the mirror is finally ground to its correct shape and the polishing
is done, it has a curve which is part of a sphere. This is what is meant by
a spherical mirror. If we want the telescope to be shorter, the curve of
the mirror has to be slightly deepened in the centre to produce the shape
known as a paraboloid, The process of changing the curve from the
sphere to the paraboloid is known as 'figuring' and is one that needs
very great care since it is very easy to over-shoot the required depth and

43

produce the fault called over-correction, thus making a slightly different curve, the hyperboloid.

A shorter telescope has some advantages when it comes to constructing a tube and a mount. This may be important in the case of a fairly large mirror. An eight-inch mirror, for instance, is quite easy to use if it is no more than about eight times longer than its diameter, i.e. a little over five feet. If the same mirror is made to work at f/12 it will mean that its focal length is eight feet. The eyepiece will probably be about that distance, or even more, above the ground when the telescope is pointing in an overhead direction. The whole thing will be much heavier and more difficult to mount.

On the other hand, the four-inch mirror of the same focal ratio, f/12, will be only four feet long and the spherical shape will be quite satisfactory in use, giving good, sharp images. There is one other advantage to the long focus telescope. It can be used with the cheapest kind of eyepieces. The Huyghenian eyepiece is one of these. Another is the Ramsden eyepiece. A short-focus mirror will need eyepieces which are of a rather special and expensive kind, such as the orthoscopic and the monocentric.

This chapter contains merely the barest outline of the mirror-making process which is capable of an infinite variety of refinements and variations in detail. Some makers, particularly in America, have been known to become so caught up and fascinated by its subtleties that the mirror becomes an end in itself and seldom, if ever, confronts a real star.

The completed mirror, in fact, is really no more than a support for a thin layer of silver or aluminuim which provides the reflecting surface. Silvering can be done by the amateur and directions for doing this will be found in most of the more advanced manuals of telescope-making, some of which you will find listed in the bibliography at the end of the book. Unless the amateur is an uncompromising 'do-it-yourself' enthusiast, it will be more satisfactory in the long run to have the mirror aluminised by one of the firms which specialise in this work. Such firms are also listed in the appendices.

Assuming the mirror is accurate enough to form a reasonable image, we now have to get near enough to it with our eyepiece to magnify it. Sir William Herschel, the famous eighteenth-century telescope-maker and astronomer and discoverer of the planet Uranus, used to tilt the mirror of his telescope to bring the image near to the edge of the open end of the tube. This is the so-called Herschelian form of reflector. It had, unfortunately, many faults, and was adopted by Herschel in order to avoid the extra loss of light which is entailed by a second reflection at the small flat mirror used by the Newtonian. Since Herschel's day, however, the coated glass mirrors have greatly improved over the reflective qualities of the old mirrors made of what was called speculum metal, an alloy of tin and copper. The small flat mirror directs the light to the side of the tube without any serious loss of light. Making a perfectly flat surface, in the optical sense, is even more difficult than making the curved surface of the main mirror and most amateurs prefer to buy the small, eliptical flat from an optician. For a four-inch mirror, of the focal length we have in mind, the ellipse should have a minor axis, or width, of near one inch. It must be large enough to intercept all the converging rays from the main mirror but not more, since it will unnecessarily obstruct the path of the light. A substitute for the flat mirror is a right-angled glass prism but, like the flat, it must be made with great accuracy to be effective and prisms of the right quality are not always readily obtainable.

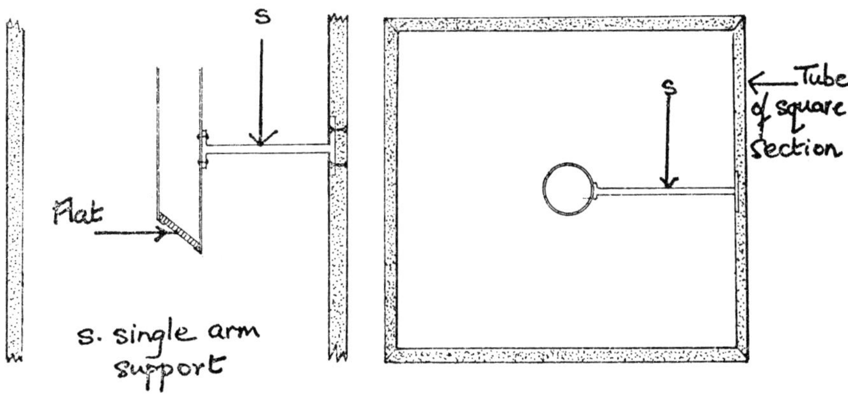

FIG 16 *Method of supporting the flat mirror*

The possible ways of mounting the mirror for use are very varied. Many kinds of cell have been used, including old brake-drums from car wheels. It is in fact doubtful whether a cell, as such, is necessary. I have used a twelve-inch mirror, as well as smaller ones, in a square wooden tube made of plywood and of open section (as shown in the diagram) with nothing more than three or four wood or rubber blocks to hold the mirror lightly in place on the base-board of the tube. Three screws, equally-spaced at 120-degree intervals, an inch or two from the circumference of the mirror and inserted through the baseboard, enabled the mirror to be tilted to the correct 'squaring-on' position. The flat mirror, too, must be adjustable and there are several ways of achieving this. Some are provided with a cell attached to the sides of the tube by a four-armed 'spider'. Others are mounted on a stronger, single arm, as shown in fig 16. Sometimes a three-armed mount is used, but possibly the best form is one which consists of two interlocking circles, made of brass or some other flexible strip of metal as you can see in fig 17. This not only holds the flat mirror firmly in place but has another important advantage in reducing what are called 'diffraction effects' which tend to detract to some extent from the image quality of the reflector, especially where stars and planets are concerned. Flat mounts of this kind can be

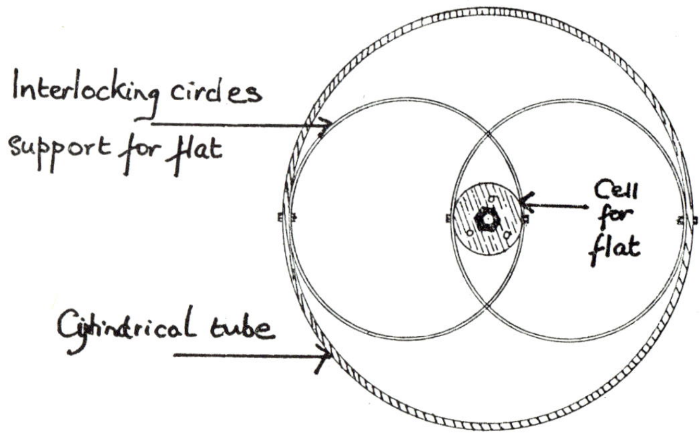

FIG 17 *Method of supporting the flat mirror*

made by the metal worker but are probably best bought from a commercial source.

The same applies to the telescope eyepiece and the focusing mount, which may be of the simple, sliding draw-tube type or the more elaborate rack and pinion kind, though the latter are apt to be costly.

It is possible to use a Newtonian reflector, especially if it has a square, wooden tube of fairly light weight, by propping it against some firm support, such as a low wall, fence or the back of a garden seat. This may sound very improbable to anyone who has been used to a rigid and smoothly-working mounting with a mechanical or electrical drive. It can be done, however, and the problem of following an object if the power is not too high, will not be insuperable, while a comfortable, downward angle of view can be obtained by turning the tube so that the eyepiece is at the top. A useful accessory, in any case, is the addition of a small finder telescope, as with the refractor, and there is much to be said for the ex-army elbow-shaped prismatic type with its very wide-field eyepiece. If it is mounted near the open end of the tube, as in fig 18, the two eyepieces will be conveniently close together.

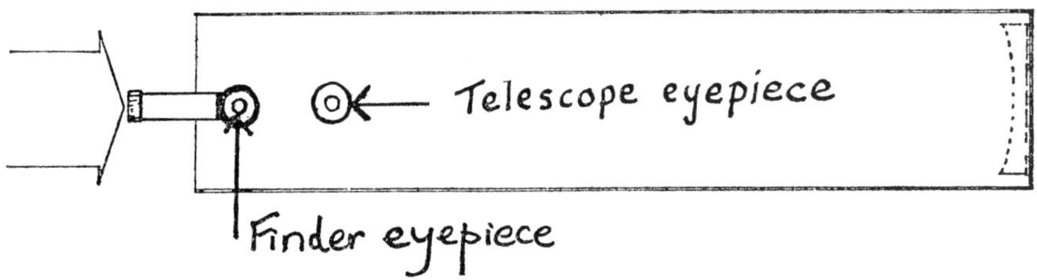

FIG 18 *Elbow-shaped finder telescope for reflector*

Collimation of the Mirrors

Any optical instrument must have its various parts put together in the right way; the lenses and mirrors fixed as firmly as possible in their correct relative positions. By this we mean that the telescope mirrors must be properly collimated and, for a good performance, the collimation must be as accurately done as possible. The main mirror, for example, must be 'squared-on' to the axis of the tube and the flat set at the exact forty-five degree angle to direct the light to the eyepiece, which, in turn, must be set at right angles to the tube and the axis of the mirror.

The first operation is to adjust the flat mirror by its screws so that the elliptical outline is tilted to present a circular shape in the eyepiece mount, the eyepiece itself being removed from the draw-tube and the outline of the main mirror concentric in the circle of the flat. A dark spot, which will be seen somewhere in the reflected image of the main mirror, is the image of the flat and this must be brought to the centre of the circle. To do this the screws which tilt the main mirror and its cell, if it has one, must be carefully adjusted until the image of the flat is brought to the required central position. A little experimenting with the screws will show which way the main mirror must be tilted to arrive at the correct collimation. A good mirror or object-glass should come in and out of focus quite sharply, with only a slight movement of the eyepiece in its draw-tube. A mirror or object-glass which is not of a very accurate shape will allow a wider range of movement without a critical point of sharp focus.

6 Making the Mount for a Reflector

Anyone who has used a small hand telescope, or even a pair of binoculars, with a magnification of more than eight or ten times, quickly recognises the difficulty of keeping the instrument steady enough to use effectively. Many binoculars now have a threaded socket or tripod 'bush', intended to overcome the snag by enabling some kind of mounting to be attached. To employ usefully a telescope with a magnification of thirty or more times, a tripod or pillar support becomes essential.

The telescope mount has two functions. One is to enable the tube to be pointed in any desired direction in such a way as to leave the eye-piece reasonably accessible. The other is to hold it firmly in position when it is so pointed, avoiding as much as possible any tremors or vibrations which will casue the image to dance wildly. Mounts fall broadly into two main types, each of which has its own good and bad points. They are the altazimuth and equatorial types.

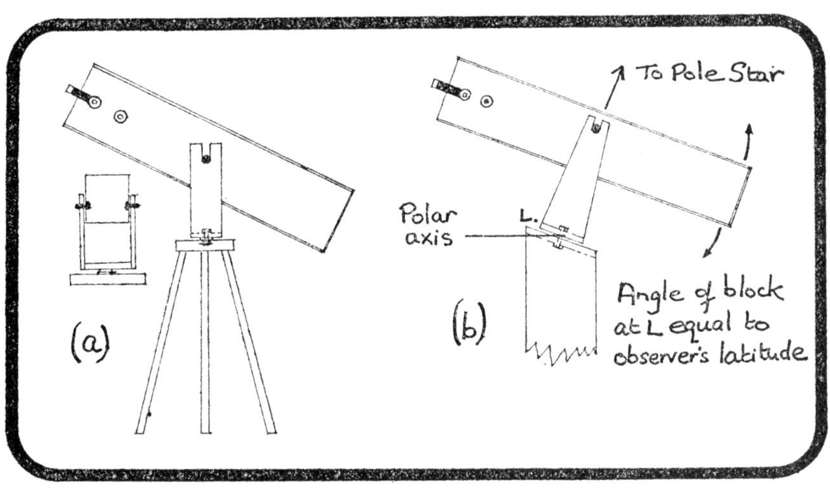

FIG 19 *Two reflector mountings*
(a) *Altazimuth tripod*   (b) *Equatorial on pillar*

The altazimuth mounting, as shown in fig 19 (a), is the one commonly employed for the ordinary kind of look-out telescope which is to be found on every seaside promenade. Its name is a contraction of 'altitude' and 'azimuth', indicating the two planes in which the telescope must be allowed to move: vertically and horizontally. The equatorial which you will see in fig 19 (b) is the type which is invariably used with large telescopes, especially those intended, as most of them are, for astronomical work. This mount has its main axis fixed parallel to the earth's axis and the telescope, once fixed at the altitude appropriate to the star or planet, will follow by one movement the constantly-changing position of the object as it makes its apparent motion from east to west, due to the earth's rotation, and much magnified in speed to the observer at the eyepiece. Large telescopes normally have some form of mechanical or electrical drive fitted to the mount so that the motion takes place without need of any adjustment by the observer. This is the ideal arrangement and, particularly for photographic work which demands that the object be kept steadily in the small field of the eyepiece for minutes or even hours at a time, it is the only effective means. The equatorial, however, is not without its weaker aspects. It really demands a fixed mount, since it must be accurately placed with respect to the observer's meridian and latitude.

For a portable instrument, such as many amateurs are concerned with, it has little advantage if not very accurately adjusted. With a Newtonian reflector there is the additional trouble caused by the eyepiece sometimes getting into an awkward position. The altazimuth, on the other hand, though requiring two movements to follow the object, can be quickly placed in any position and, in the case of the Newtonian reflector, presents the eyepiece always in a horizontal position, which is an undoubted convenience.

From experience with both types of mount the writer has found that the altazimuth has much to commend it from the amateur's point of view. Some types of equatorial mounting, notable the so-called German form, have the telescope tube fixed at the end of the declination axis. This is very awkward for a portable telescope and it demands a heavy counterweight to balance the telescope-tube if the latter is anything but short and made of very light material.

7   Using the Telescope

There is a saying among amateur astronomers – and it probably applies with equal force to professional ones – that the most important part of the telescope is 'the man at the small end'. Certainly it should never be overlooked that the observer's eye is a vital part of the instrument. One observer will see with a two-inch telescope what another may miss with a ten-inch one. Let us, for a moment, consider what the smallest telescope with only a one-inch lens is capable of. First of all it will collect about twenty-five times the amount of light which will normally pass into the average human eye with the pupil open to a diameter of one fifth of an inch. On a favourable evening, stars as faint as the ninth magnitude may be seen in a dark sky, many thousands more than the number seen without it. The author of that famous old classic of amateur astronomy, *Celestial Objects for Common Telescopes*, which was first published over a century ago, remarked that "even diminutive glasses if good, are not to be despised. They will show something never seen without them."

Some fine double stars can be seen if the telescope is steadily supported. The separation of these will reach to 4·5 seconds of arc, which is equivalent to showing as two points of light, street lamps about a dozen yards apart at a distance of over two hundred miles. If the eyepiece gives a magnification of about 40x, the one-inch telescope will show the planet Jupiter about the same size as the full moon to the unaided eye, although this may be hard to believe. The four largest satellites will also be seen and their revolution followed from night to night.

Looking at the Sun

Never use any telescope, or even binoculars, to look directly at the sun unless, possibly, on those rare occasions when its light is reduced by mist or haze and even then it is important to use a dark screen to reduce the light still further. It is better to use the telescope as a projector and look for sunspots on the sun's image cast on a white screen of paper or card held at some distance from the eyepiece and focused as sharply as possible. Sunspots are visible on most days though their number varies considerably. They also vary in size. The discovery of the sunspot cycle of about eleven years is a good example of what may be achieved with very modest instruments in the hands of a patient and persistent observer.

The rise and fall of the cycle was discovered by a German amateur astronomer named Schwabe, of Dessau. He lived just over a century ago and was anxious to find a planet which was at one time believed to be revolving round the sun inside the orbit of Mercury. He reckoned that such a planet, (which had actually been given the appropriate name of Vulcan, after the ancient god of fire) would probably be seen at some time passing across the face of the sun, just as Mercury itself, and more rarely, Venus are, when passing between the earth and the sun. Schwabe patiently watched the sun on every possible occasion over a period of many years. The planet Vulcan eluded him but when he came to study his careful sunspot records he found the eleven-year rise and fall which is now always associated with his name.

Looking at the Moon

The moon presents an attractive target for the smallest telescope, or even a pair of binoculars. At full phase it is an object of great brilliance although, surprisingly, the surface is not a very efficient reflector of the sunlight which illuminates its landscape. Not more than about seven per cent of the sun's rays are reflected back from it. The moon's 'day' lasts for two of our weeks and, at full moon, the sun is high in the lunar sky and the sharply-etched shadows, which throw the mountain ranges and craters into clear relief at the intermediate phases, are absent. Nevertheless, the full moon enables us to identify many of the main lunar features, particularly the darker areas, the 'maria' or so-called lunar seas. The larger ones, of course, are clearly visible to the naked eye, but if we are using a map to identify them by name, it is as well to note that most maps show the moon as an astronomical telescope does, in an in- verted aspect. Binoculars, or a telescope with an erecting or 'terrestrial' eyepiece, will make it necessary to turn the map round the other way.

Perhaps the most interesting feature presented by the full moon are the mysterious light rays which surround some of the craters, notably the crater called Tycho. These make a telescopic view of the full moon appear something like a peeled orange. The crater Copernicus is another centre of these rays while the smaller crater, Kepler, shows them too. Here and there across the dark regions of the lunar 'seas' we may detect the signs of old craters, mere ghosts of pits now submerged under dust or lava flows which form the surface of the maria.

However interesting the face of the full moon, the real nature of our satellite can be better seen at the phases in between new moon and full, or afterwards in the later hours of the night and early morning, when the moon is waning. This is when the slanting rays of sunlight cast black shadows from the crater walls and mountains along the line of the terminator as it is called, where the sun is rising or setting. Here, even small craters and mountains become conspicuous objects, while in the full phase much larger ones quite lose their identity and many disappear from view. With the help of a map, such as that originally drawn by Elger and later revised by another amateur expert, H. P. Wilkins, it is surprising how quickly the beginner will learn to find his way about the moon and become familiar with the grand lunar spectacles, such as the sun rising on the terraced walls and central mountain peaks of the Copernicus crater, or the vast walled plain of Clavius in the region of the moon's south pole.

Other phenomena of interest to the observer of the moon are the occasional occultations of stars or, more rarely, of planets. When the moon is between new and full, it is the dark edge which is leading in its real west to east motion, and it is this movement which takes the moon in front of a star, so that the star is first hidden at the invisible edge. It is quite astonishing to watch the gap between the moon and the star narrowing and then to see the star disappear quite suddenly, as though it had been turned off at a switch. This indicates quite clearly that the moon has no appreciable atmosphere. If it had any the star would be seen gradually fading before disappearing.

Mercury

Five of the sun's planets have been familiar to sky-watchers from ancient times, long before the invention of the telescope. It must have needed fairly close observation to detect the smallest planet, Mercury, which never moves far from the sun and is only seen for a few evenings in the spring when near its greatest apparent distance, called elongation. This is when Mercury lies to the east side of the sun. At its western elongation it appears most clearly in the east, just before dawn, for a few days in autumn. This applies to observers in the earth's northern hemisphere. In the southern hemisphere it is more conveniently placed as a morning object during the spring and as an evening 'star' during autumn.

Mercury's motion is so rapid that it appears three times a year in the morning and three times in the evening, but can only be well seen at the times mentioned above, the exact dates varying. At other elongations it will be too near the horizon by the time the sky is dark enough in the evenings. In the eastern morning sky the daylight will be too bright by the time it gets clear of the horizon. In fact, at a favourable time, Mercury is quite a bright object, equal in brilliance to some of the brightest stars.

It is a satisfying experience to see Mercury at all, 'shining' as one American astronomer says 'like a globule of molten metal' against the sunset sky. A small telescope may enable it to be seen sometimes before it becomes visible to the naked eye and, with sufficient magnification, will show its phases like a small moon. Even larger telescopes are seldom really effective with this still little-known planet, and although it has no atmosphere so that the solid surface is exposed, very little in the way of surface markings has ever been satisfactorily seen.

Venus

Venus, which revolves outside the path of Mercury, is a very different kind of planet. When Venus is above the horizon in a dark sky there is no mistaking it, for it far outshines everything else but the moon. Indeed, the splendour it reveals as a naked-eye object may lead us to suppose the telescope to show something even more spectacular. Most telescope-viewers, however, find Venus rather a disappointment. The highly-reflective surface is due to the dense, cloud-laden atmosphere which surrounds the globe, and the bright rays of the planet are a severe test for any telescope, particularly against a dark sky. The telescopic observer will find it better to look for Venus against the twilight heavens, when its lustre is less overpowering.

The crescent phases will easily be seen, much as they were shown by Galileo's telescope three centuries ago, but hopeful generations of astronomers have yet to see a gap in the cloud layer which will reveal its surface. From time to time there have been reported dusky shadings of one kind or another, and some optimistic observers have even claimed to have detected the existence of mountains, but so far the telescope alone has proved unable to do much more than show us the phases. Nevertheless the shimmering crescent of brilliant white is a sight long to be remembered.

A magnification of no more than about twenty times is enough to show the crescent phase but, when approaching the gibbous phase, like a nearly-full moon, the planet is farther away, much smaller and needs a higher magnification. Transits of Venus, which occur when the planet's movement takes it exactly between the earth and the sun, are extremely rare, the next one being due to occur in the year 2004, followed by another one eight years later. After that there is a long gap of a century or more. When a transit does occur the planet is seen as a black spot passing across the face of the sun when the bright rays are screened by dark glass.

The Red Planet

Mars has perhaps caught the imagination of most people. When it makes a fairly near approach to the earth it shines with a bright, red light and, being an outer planet, it may come high into the southern sky and well away from the direction of the sun. The small telescope will provide an interesting and instructive view provided the beginner has the patience to educate his eye, and does not expect to see a globe as big as the full moon streaked with the famous 'canals'. Mars has no dense atmosphere like Venus and we can see the solid surface but, at even its closest approach, just under 35 million miles, it is still a small object and needs somewhat higher powers than our two-inch refractor will provide. At a favourable time, however, the most prominent of the darker markings, the one named Syrtis Major (or the Hour-glass Sea) has been sighted with a two-inch telescope. The white polar caps are an attractive feature, contrasting strongly with the predominantly orange-red globe and the blue-green streaks and patches. Perhaps the most interesting revelation of the recent space-probes to Mars has been the discovery of craters on the surface, closely resembling those which abound on the moon. They were previously quite unsuspected since they are far too small, at a distance of many millions of miles, for even the largest of telescopes. Mars, more than most planets, shows the limitations of the small telescope, demanding fairly high magnifications and good viewing conditions in the atmosphere.

Jupiter, the Giant World

The multitude of tiny planetary bodies known as the asteroids, revolving round their paths between the orbits of Mars and Jupiter, are of little interest to the user of a small telescope. Beyond them the largest planet of the solar system provides the most rewarding object in the sky, apart from the moon. With a magnifying power of about forty times, Jupiter's giant globe will be seen near opposition as big as the full moon is to the naked eye. A good two-inch object glass with a magnifying power of slightly more than this, such as sixty times, will make a fine spectacle of Jupiter and its four largest satellites, Io, Europa, Ganymede and Callisto.

Every year Jupiter shines brilliantly in the night sky, for several months at a stretch. The best time to examine it is when it is near opposition, when the earth lies on the line between Jupiter and the sun. The planet is then at its nearest and its main features will reveal themselves more readily. The globe will be seen bulging round the equator and flattened at the polar regions.

The solid surface of Jupiter has never been seen and we have no exact knowledge of what lies beneath the dense atmosphere composed of hydrogen, methane and ammonia. Nevertheless, the upper layers of the atmosphere display much of interest to the owner of a small telescope. The main cloud belts can be well seen with a two-inch refractor and the six-inch reflector will enable much detail to be detected in their changing

56

forms. The beginner is unlikely to see the unique feature of Jupiter, which is known as the Great Red Spot, unless his telescope has an aperture of about three inches and a magnification of 100x or more. But I well recall seeing the tiny black shadow of a satellite crossing the bright globe, with a telescope of little more than two inches in diameter made by the famous old English firm of Thomas Cooke, whose excellent object-glasses had a well-deserved reputation among both amateur and professional astronomers all over the world.

The constantly-changing details of the cloud-belts need a good three-inch refractor or a six-inch reflector, and the records of the Jupiter section of the British Astronomical Association contain ample evidence of the good observations which have been achieved with such small instruments in careful and patient hands.

The Ringed Planet, Saturn

The furthest of the sun's planets known in ancient times is probably the most spectacular object in the solar system because of the beautiful system of rings which surrounds it. The observer who has not concentrated too much upon the illustrations, whether drawings or photographs, made with the aid of very big telescopes, will be well-pleased with the sight of Saturn even in a two-inch. I have very vivid recollections of my own first glimpse of the planet, high in the autumn sky, with the rings near their wide-open phase. The telescope was only two inches in aperture and fitted with an eye-piece magnifying about sixty times. Larger ones, such as the six-inch reflector or anything bigger still, will present the observer with a memorable view. The largest satellite, Titan, will be seen in a two-inch telescope and another inch of aperture will show two or three others on a clear night. The rest of the ten Saturnian moons need a larger telescope and a keen eye. The main division in the rings, named after Cassini who first saw it, is visible in a three-inch when the rings are well opened, though it has been claimed that this, and even the shadow cast by the globe of Saturn on the rings, can be seen in a two-inch telescope. It should be remembered that details of this kind, which may appear quite invisible to the beginner, may well have been seen by someone with a very experienced eye. It is significant that when you have looked at an object with a large, or moderate-sized telescope and your eye has become familiar with it, a great deal of what has been seen may still be visible with a much smaller aperture.

The other worlds of the solar system cannot be said to present a picture of much interest with small apertures. Uranus can be seen since it is just within reach of the naked eye when near opposition and when the atmosphere is very clear. Even in a large telescope the very small, blue-green disc of Uranus is rather a remote and featureless spectacle, while Neptune and Pluto lie far beyond reach of the naked eye. Although Neptune can be seen with a small telescope it is difficult to distinguish it from a faint star, but a six-inch reflector will show a very minute disc. Pluto is a very small world as well as the most distant one, and nothing less than a telescope of about ten-inches aperture will make it visible.

Looking at the Stars

Although the stars are incandescent globes of gas, no telescope is capable of showing a star as a true disc. There is a tiny disc visible, surrounded by one or two narrow rings of light, but this is what astronomers call the 'spurious disc' and oddly, perhaps, the larger the telescope, the smaller this disc appears. To the expert eye the disc and rings at the focus, or the larger disc into which the star expends inside and outside the focal point, are capable of telling most of what there is to know about the quality of the telescope's object-glass or mirror. Much, of course, depends on the state of the atmosphere. A night when viewing is bad will make it impossible even to detect the spurious disc, and the star will appear as a shapeless, agitated blob of light. Covering part of the object-glass or mirror ('stopping-down') often helps, since there is then less turbulence in the smaller column of air through which the light passes. The object-glass, particularly of a small refractor, normally shows a sharper and cleaner star-image than a reflector, which is not only, as a rule, of a larger aperture but is liable to some further interference from the presence of the small, flat mirror with its three or four arms supporting the mount, and because the light has to travel twice along the tube.

A night which is exceptionally clear may not always produce the best conditions for observation of either stars or planets, although it may be very satisfactory for picking up faint objects or star-clusters which normally require a low-power and a wide field of view.

Stars are classified in relative brightness by a series of numbered 'magnitudes', the first magnitude containing about twenty of the most brilliant. Succeeding magnitudes, denoted by a higher number, each show a marked increase in the number of stars, as would be expected since each step includes a larger sphere of fainter stars and, as a general rule, the bright stars are likely to be among the nearer ones. Sirius, the brilliant leader of Canis Major, the constellation of the Great Dog, is an obvious example. It is the brightest of all the stars and its distance from us is a little under nine light-years, a relatively near neighbour of the sun.

In looking at any celestial object, it is advisable to begin observation with a low power instrument. There is little advantage to be gained by using a high power to study what is seen satisfactorily with a lower one. The high power will show a smaller field of view, it will produce an image which is much fainter, though larger, and it will magnify both the speed of movement (due to the earth's rotation) and the ever-present turbulence of the atmosphere which, in its worst form, produces what astronomers know as 'bad seeing'. This is accentuated, too, in telescopes of larger sizes which collect a larger column of light and this is more liable to disturbance than the narrower rays collected by the small object-glass or mirror.

The night sky abounds with attractive examples of double, triple and larger groups of stars, as well as the mysterious gas clouds known as nebulae and the distant galaxies which often resemble the nebulae in appearance. There are many books which contain useful lists of all the constellations visible throughout the year wherever you may live. Some of these books you will find in the bibliography. It is important to know the main outlines of the constellations, of course, and no optical equipment is needed for this. Even the ability to identify the brighter star-groups in their seasons through the year will give very great satisfaction. There are a number of star-maps and simple atlases available to assist the beginner in this pleasant task. The revolving star-map known as a planisphere is a valuable aid in this since it can be turned to show the aspect of the sky at any hour of the day or night throughout the year.

8   Other Telescopes

It is quite natural to become more ambitious and telescope-makers are very much like other people in this respect. After using a small telescope you may find yourself wondering how to progress to something bigger and better – though never forget that 'bigger' need not always mean 'better' in spite of the saying that a good big one will always beat a good little one. So far as telescopes are concerned this is not always true. Nevertheless, the fact remains that many people who have made a small telescope do go on to make bigger ones.

So far as the refractor is concerned, you will find that for anything beyond about two inches in diameter the simple convex lens is not of very much use. The rainbow colours it produces become very troublesome. This was one of the things that bothered the early makers of refractors and so they tried to overcome the problem by making their telescopes very long, even fifty yards and more. A larger refractor of three, four or five inches, needs to have a very good quality lens and these become extremely costly. Everything else, such as the tube and mounting, must be corrsepondingly large and heavy. Carrying a five-inch refractor outside on a dark night and lifting it on its mount, which might be a tripod of seven feet in height, is no easy matter. A refractor of this size really needs a permanent mount and, unless you are fortunate in having a clear space for some distance all round you, there will be times when your target will disappear, for hours on end, behind chimneys or trees.

A refractor of three or more inches in diameter will be quite expensive. Most amateurs who look for a bigger telescope find that the choice has to be the Newtonian reflector of eight, ten or twelve inches. Mirrors of this size need not cost very much to make although they do involve a lot more time and trouble. They can be bought fairly readily from many of the optical suppliers who deal in telescopes and an eight-inch mirror with its small elliptical flat should be no more than the cost of a bicycle and may be found second-hand for rather less. Mounted in a square wooden tube it will give a very good performance and is not too heavy to carry outside from a garage or garden shed.

The Newtonian reflector is not the only kind of mirror-telescope, though it is certainly the simplest to make. There are two quite old designs called the Gregorian and the Cassegrainian. These are rather like short refractors in their outward appearance and they are used in much the same way, since their eyepiece is normally at the bottom end of the tube. They have a certain advantage in being short and compact in size but making the mirrors for them is somewhat more difficult than it is in the case of the Newtonian.

The Gregorian telescope with a short brass tube and mounted on a small table-tripod is one you may sometimes find in antique shops. The Cassegrain telescope is still made but costs more than the Newtonian with a similar-sized mirror. Apart from its greater cost, it also has some other disadvantages and has sometimes been said to combine the bad points of both the Newtonian and the refractor. It is, however, the design used for many of the world's largest telescopes.

The Mirror-lens Telescopes

Every reflecting telescope is, in one sense, a combination of mirror and lens, since all use an eyepiece, at least for visual observation if not always for photography. In recent years a new family of telescopes has been devised and, in one form or another, they may well become the instruments of the future. They are sometimes known as catadioptric telescopes and one of the best-known of them is the Maksutov, named after its Russian designer. It is, in fact, another modification of the Cassegrain design, but it has some advantages in that its optical parts are all of spherical shape. They must, however, be very accurately matched and, the advantage of the short, closed tube, is achieved by the use of a kind of lens called the correcting-plate, which, like the refractor's object-glass, must be of high-quality optical glass. There is no supporting arm for the small secondary mirror which can be supplied by a small alumised spot on the correcting plate.

The great advantage of the Maksutov design is its compactness. It can provide an effective focal length of eight feet in a tube less than two feet long. It has few optical faults and its closed tube is unaffected by the turbulent air currents which often affect the open tube of the Newtonian. Undoubtedly the Maksutov has an assured future, though hardly yet for the home-constructor. It is decidedly not a telescope which can be made like the Newtonian, on the kitchen table, although, unlike the Newtonian, it can easily stand on one and often does.

9 Observing with Binoculars

However big a telescope you have, there is always a place for a pair of binoculars in your observing equipment. Many people use them on holiday or for watching sport, but seldom turn them to the sky on a clear night. Of course they have a low power of magnification but there are times when this is an advantage. Binoculars are very much smaller than most telescopes; they have a wider field of view and looking at things with both eyes, as we normally do, is undoubtedly the most comfortable way of observing. Using both eyes we see things from two slightly different positions and this gives us the appearance of depth which we call stereoscopic vision. We do not get this effect by using one eye. The two tubes of the binocular are spaced more widely than our eyes and this gives an increased impression of depth which most people find pleasing. Fig 20 shows how the light rays travel through a pair of prismatic binoculars.

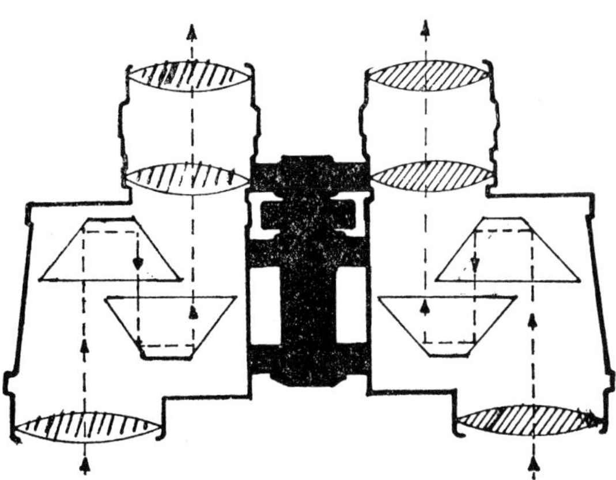

FIG 20 *Prismatic binoculars showing light-path*

Binoculars are really two small telescopes mounted side by side. There are two kinds. The best ones are prismatic, as mentioned above, which means that each tube contains two prisms for the purpose of folding the light back so that, instead of needing a long telescope tube, the whole instrument is very short and compact. There is one problem with prismatic glasses, however, and this concerns the prisms, which have to be very accurately fixed in position. If there is the smallest error, or if one prism is displaced by an accidental knock, the two images will not merge exactly. If the error is very slight, your eyes will try to correct it, but in doing so will strain your eye-muscles. There is a simple test which you can use to tell whether the lenses and prisms are exactly right. This is what is meant by collimation and means the same as it does in the case of the mirror adjustments in a reflecting telescope.

Focus your binoculars on any horizontal line, such as a roof ridge across the street, or a telegraph wire. Move the binoculars a few inches away from your eyes. You should now see the two fields of view separated by a space between the tubes. The important thing is that the two short lines, though broken, should be still in line with each other. If you open or close the hinge which allows the binoculars to be adjusted to the width of your eyes, the two images will move apart or come closer together, but the two lengths or wire or sections of roof ridge should still be in line. In many cases they will not be. Do not use any pair which fails to pass this simple test. If you do, you risk damage to your eyes. Spectacles have to be sold by qualified opticians to ensure their accuracy and suitability but anyone can sell binoculars and a faulty pair can do nearly as much harm as unsuitable spectacles.

Some binoculars are of the kind often known as Galilean glasses because they are similar in design to the telescope used by Galileo. The ordinary opera-glass is one of these, but there were others made before the prismatic type was invented. Some are still manufactured and they are always much cheaper than the prismatic ones though they have the disadvantage of a rather small field of view and the powers of magnification are somewhat lower, usually not more than three or four times. Even so, they will give pleasant views of the night sky and a bright, sharp image. Some years ago, in fact, an entire book was written on the opera glass and its use for observation of the stars. It is not often seen now but you may come across it in a library or a second-hand bookshop. The title is *Astronomy with an Opera-glass* by an American author, G. P.

Serviss. Another more recent book on astronomical work with binoculars is listed in the bibliography while the present writer's *Night Skies of the Year* also contains references to the subject.

There are many binoculars of good quality to be obtained from dealers in surplus government equipment. These are almost always either 6×30 or 7×50 glasses. The first figure in each case is the magnification of the binocular and the second is the diameter of the object-glass, in millimetres. The 7×50 glass is somewhat larger and usually costs more than the small 6×30 one. Because of its larger object-glass it gives a brighter image, but not always a wider field of view. In fact the 7×50 binoculars are fairly close to the spceification of the 'rich-field' telescope and will give splendid views of the Milky Way and star clusters.

Most modern lenses are coated with a substance which reduces the loss of light by reflection at the surface of the lenses and prisms. This coating, sometimes called 'blooming', lets more light pass through the lenses and shows itself on the outside of the lens by a coloured reflection of blue, purple or amber, depending on the material used.

It is always a good idea to begin by looking through the objective end of binoculars before you buy them. If you do this while holding them up to a good light you will see at once if there are any edge chips on the prisms and any dust or dirt on them. This may well be seen if they are old or second-hand ones. Very old or neglected ones which have been lying in a damp place may even show a growth of a fungus on the prisms! Leave these alone, for cleaning or dismantling binoculars is decidedly a job for an expert and may be expensive. Incidentally, do not be deceived by the views which are supposed to be seen through binoculars as shown in films and television plays. These almost always show a field of view shaped like two interlocking circles. The truth is that no properly adjusted binocular shows a field which is anything but a perfect circle.

You may come across the small prismatic telescope which is often called a monocular. These are very useful and always much cheaper than a binocular of equal power and size. They give exactly the same magnifying power as the equivalent binocular and also the same field of view. This is to say that a 7×50 monocular is just the same in these two respects as a 7×50 binocular. The only thing which differs so far as their performance is concerned is that the monocular does not give the same impression of depth, the stereoscopic effect, which is obtained by the use of both eyes in the case of binoculars. Against this slight disadvantage the monocular has one very good point in its favour. It is far less likely to be troubled by faults of collimation in the prisms which are so common with binoculars where the two images must be brought very accurately together.

The magnifying power of these small prismatic telescopes, whether of the binocular or monocular kind, is not enough to show much on the planets. You can, however, see the four large satellites of the planet Jupiter and, if the magnification is eight or more times, you can just see Saturn's rings. On the moon it will show many of the larger craters and some of the mountain ranges. You may be interested to know that a book has been written entirely on the subject of the moon as it can be seen with such a small instrument. The field for which the binocular is particularly suited, however, is that of the stars. Some expert workers in astronomy make use of binoculars for the study of variable stars. These are stars whose light changes over a period of time which may be a matter of months, days or even a few hours. During the last war some very large and heavy binoculars were made for use on ships and anti-aircraft sites. You can still find these for sale although they are rather expensive. These have been used by amateur astronomers who spend many hours of their leisure time searching for comets. Here, the important thing is a wide field of view and fairly low magnifying power, enabling a faint comet to be spotted when it is no more than a misty speck in the sky and millions of miles distant. Using binoculars, too, you will often catch sight of a shooting star, or meteor, as it flashes across the field of view. It will show many which are quite invisible to the naked eye and there are large numbers of these which rush into the earth's atmosphere and shine briefly before the heat of friction burns them up.

The binocular gives particularly attractive views of some of the bright star-clusters, such as the Pleiades and Hyades, and will clearly show some of the more prominent nebulae and galaxies such as the Great Nebula in Orion and the famous Andromeda Galaxy. Of these and the many other objects of interest in the sky you can learn from some of the books listed in the bibliography, and some of these contain star-maps which show you where to find them among the constellations throughout the year.

10 The Sun Clock and Star Clock

For thousands of years before the invention of the mechanical clock our ancestors had nothing with which to measure time but their observations of the sun and stars. Our intervals of time, the day, month and year of course, are still based upon the movement of the earth round its axis and its annual journey round the sun. Our calendar, too, has its beginning in the same way. Although we no longer rely upon the sun and stars to tell us the time or the season of the year it is still interesting to make and use the simple instruments which measure the passage of time up to three or four hundred years ago. This chapter tells you how they are made. The sun clock was the instrument we now call the sun-dial. The star clock is less often seen outside a museum. Another name for it is the nocturnal.

Making the Sun-dial

The materials needed are:
1. an eight or nine-inch square of plywood
2. a sheet of rigid cardboard or plywood about five inches square
3. a few miscellaneous items including a protractor, set-square, ruler and a tube of strong adhesive.

There are two parts to the sun-dial: the face or dial itself and the style, or gnomon, as the shadow-casting finger is called. Making the sun-dial requires cutting the gnomon in the right shape and fixing it in position on the dial, which is marked out in the correct divisions to record the hours as the shadow moves across it. The method of marking the dial is shown in fig 21.

First draw a line exactly down the middle of the square base thus dividing it into two equal rectangles. This will be the twelve o'clock line on the dial. Two or three inches from one end of this line draw another across the board at right angles to the first. This is the six o'clock line and the point where the two lines cross is where the gnomon is placed. This point is marked X in the diagram. From the point X a line is drawn which makes an angle with the first line equal to the latitude of

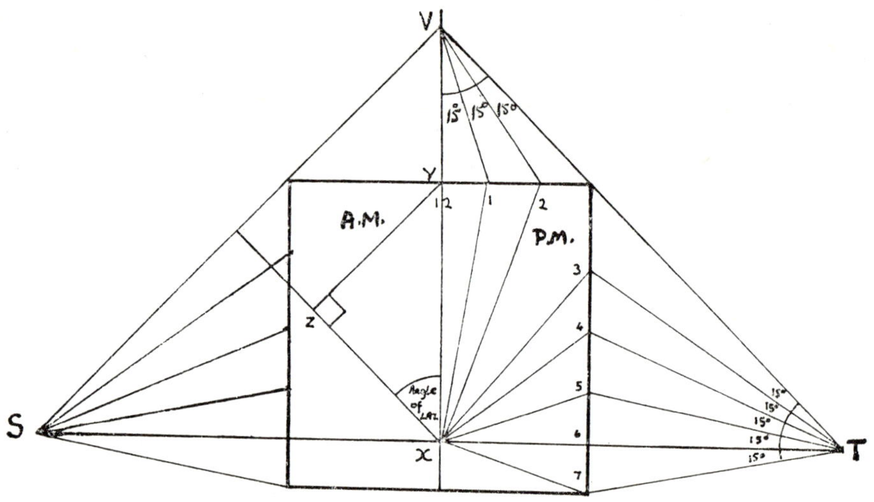

FIG 21 *Marking out the Sun-dial*

the place where you live. If you do not know this angle you will find it marked on a map. In the British Isles it will be somewhere between fifty and sixty degrees north of the equator. From the point marked Y in the diagram draw a line which makes a right angle with the line XZ. This distance, YZ is now marked out on the first line extended beyond the point Y to V. (i.e. YZ=YV). At this point, (V), we make angles of fifteen degrees on each side of the line XV until they miss the sides of the dial.

Two lines are drawn from V through the corners of the dial to meet the six o'clock line at S and T. At these two points, S and T, we also draw angles of fifteen degrees and these lines mark the edge of the dial in the hour intervals needed. The points where they meet the edges of the dial are all joined to the point X. and numbered as shown. The style, or gnomon, is cut out of cardboard or plywood in the form of a triangle with the sloping edge making an angle with the dial equal to your latitude and meeting the dial at the point X. It can be attached to the dial with glue or small tacks.

Sun-dials fixed on the walls of churches or other old buildings are designed somewhat differently. The dial described here is the more familiar horizontal one and it is often found in use as a garden ornament. To be of any practical use, it must be placed so that the gnomon is in the meridian, (mounted as shown in fig 22), the north to south line through its position, with the high point of the gnomon to the north. In fact it should be pointing towards the Pole Star. A good idea is to draw the

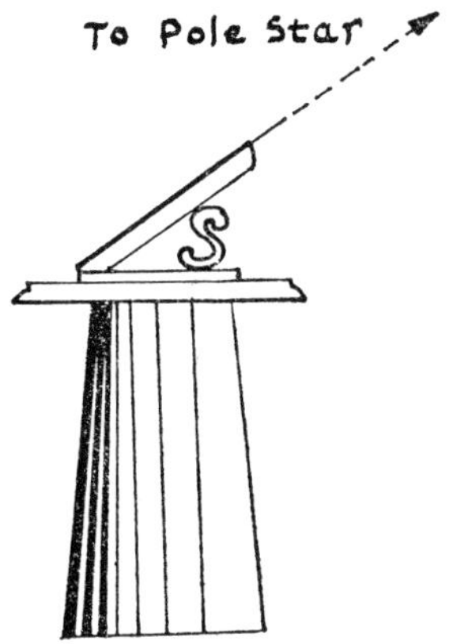

FIG 22 *The Sun-dial mounted*

plan of the dial on a sheet of paper first and then place the paper on your plywood, with a sheet of carbon paper between, and go over the lines which mark the divisions round the dial. The actual measurements of the dial are not important and the method of making it is the same whatever size we choose.

The sun-dial will tell the time by the sun only for your particular place. For places to the east, noon (or any other hour) will come a little earlier and anywhere to the west it will occur a little later, the exact difference depending on the distance away. Radio time-signals will tell Greenwich Mean Time and this may differ from your sun-dial by as much as fifteen minutes, varying according to the time of the year. The difference between sun-dial time and Greenwich Mean Time, or clock time is called the Equation of Time. The clock and the sun-dial only agree on four dates during the year.

Making a Star Clock

The sun-dial has the obvious drawback that it is quite useless when the sky is cloudy and at any time when the sun is below the horizon. Indeed, before clocks were invented a cloudy sky made time-keeping a very difficult operation since clouds hide the stars as well as the sun. Nevertheless, there are many clear nights, even in the uncertain climate of Britain, and the star-clock or nocturnal, to use its old name, then came into its own.

The east-to-west movement of the stars, like that of the sun, is an appearance produced by the earth's daily rotation and so the stars can be used at night for telling time as the sun can be by day.

In some respects the stars have an advantage, for there are star-groups, or constellations, which never sink below our horizon. To find these we must look towards the northern part of the sky. There, at a height above our horizon equal to our latitude, measured in degrees, we shall find the Pole Star. In fact the Pole Star is not exactly at the north pole in the sky, but it is very near it. The Pole Star appears to remain still, with the other stars circling about it every twenty-four hours as the earth carries us round. Since the Pole Star does not mark the position of the pole exactly it does, in fact, make a very small circle itself. But for our present purpose we can regard it as fixed.

Around the Pole Star there are a number of constellations which never set. These are called the circumpolar groups. The most familiar of them are the seven bright ones which make up the Plough or Dipper, though they are actually part of a much larger constellation called the Great Bear, or Ursa Major, to use its Latin name. The seven stars of the Plough are the hand of our clock and the sky itself is the dial. In particular we use the two stars called the Pointers which lie at the outer edge of the group, the side of the Dipper or saucepan farthest away from the handle. These stars get their name because they point roughly to the Pole Star.

We shall find them rather low down in the northern sky if we look for them in the evening during autumn. In spring, on the other hand, they are to be found high overhead, even if we look for them at the same hour.

70

The reason for this is that the earth travels round the sun a certain distance each day. Measured in miles it is a very long way for we are moving at just over 18 miles per second, but measured in degrees it is only about one degree daily. The complete circuit of the sun, which takes a year, represents a turn of 360 degrees and of this a tiny fraction under one degree is covered every day. As a result of these two movements the stars of the Plough make one complete turn round the Pole Star every day as the earth spins on its axis and a little bit more, approximately one degree, as a result of the other turning movement caused by the earth's revolution round the sun. Because of the little extra movement we find that from night to night, if we look for them at exactly the same time, they have moved a little farther in their circular track round the celestial pole. This complicates our star-clock a little, but not too much, since we can make the necessary adjustment for the date. It is, however, rather like having a clock which not only moves its hands but also has a dial which moves slowly round once in a year.

To make the star-clock we need the following: a piece of cardboard about sixteen inches long and eight inches wide, a round-headed paper fastener (not a wire clip), a pair of compasses, scissors, a knife and a cutting-board. If you are handy with a fret-saw you can make a better one by using plywood or some other material more rigid than cardboard. Some of the ones made centuries ago were of brass and beautifully engraved and decorated. These are still to be found in museums with collections of old scientific instruments.

To make the nocturnal, first cut the sheet of cardboard into two equal squares, each having sides of eight inches. From one of the two squares a circle of about three inches radius is cut out and a small 'window' hole about half an inch square, is cut in the position shown in the diagram (fig 23 (a)) between two of the radius lines which mark out angles of sixty degrees. In the sector of the circle which is diametrically opposite the window hole draw in the straight line joining the two ends of the diameters. This line will always indicate the position of your horizon. The centre of the circle shows the position of the celestial pole, marked approximately, as we have seen, by the Pole Star. The seven stars of the Dipper are drawn in each of the six sectors of the circle as shown in fig 23 (a) indicating the movement they follow in the course of every twenty-four hours.

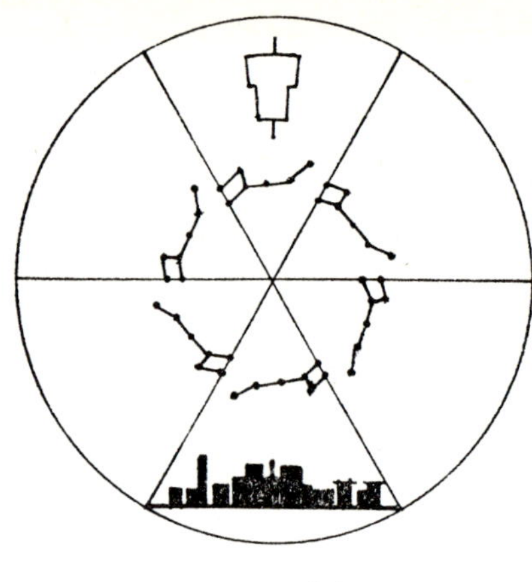

FIG 23(a)

On the other cardboard square draw the five concentric circles shown in fig 23 (b), the largest one of radius four inches and the others, proceeding inwards, of three and a half inches, three inches, two and a half inches and two inches. Draw a diameter to the outer circle and mark off angles of thirty degrees so that there are twelve spaces or sectors. On the outer circle these are marked with two-hourly divisions from twelve o'clock midnight throughout the twenty-four hours, although of course the nocturnal can only be used when it is dark enough to see the stars.

Place the small cardboard disc centrally over the large one and make a mark on the lower one through the small window hole. In the ring where the mark appears write the months of the year, the dates showing the seventh day of the month. March 7th should be marked on the line which shows midnight on the hour circle. That is the time when the two stars of the Pointers are in a vertical line with your horizon. You will see from the diagram that the months are marked from March onwards following a clockwise direction while the hours from midnight go anti-clockwise.

The smaller circle is then attached with a paper fastener, through a small central hole to the larger one so that it will revolve easily, but not too loosely, over the lower dial. This completes the making of the star-clock.

To use it, you hold it in front of you so that the centre of the dial,

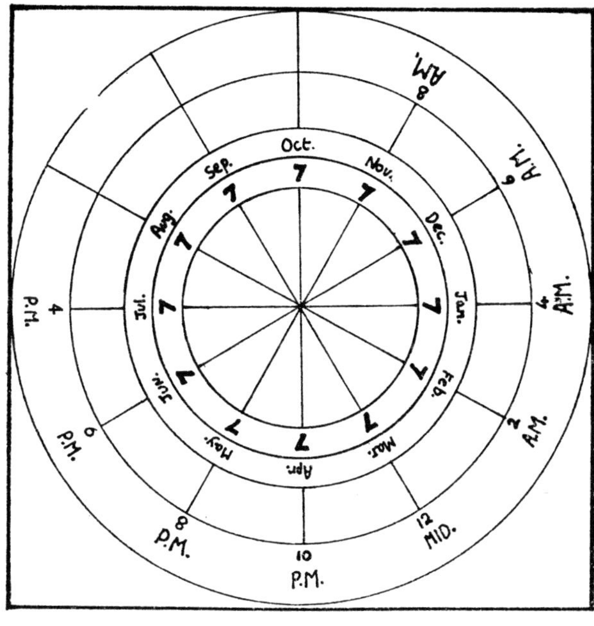

FIG 23(b)

marked by the fastener, is approximately over the Pole Star. The smaller dial is turned so that it shows the date, as nearly as possible, through the small window hole. The whole clock is then turned so that the horizon line is level at the lowest position of the circle. Look at the Dipper stars to see which of the six star diagrams coresponds most accurately to the position of the Dipper. A line from the Pointers in that section of the star diagram is followed to the outside edge of the dial and this will show you the time. By this means it should be possible to tell the time to within about ten or fifteen minutes. Of course your watch will do this more accurately, but even the best watches can run down or go wrong and it is interesting to remember that the first clocks and watches were so unreliable that a small pocket sun-dial or nocturnal was often carried in those days to check the time-keeping of the watch or clock! Apart from any practical use as a time-keeper the nocturnal, like the sun-dial, will give you a better understanding of the changing appearance of the sky through the seasons.

Another very simple and useful instrument you can make for finding your latitude is the quadrant, shown in fig 24. This consists of a quarter of a circle with a radius of about six inches cut out in wood or cardboard and marked in degrees like a protractor cut in half. A small object such as a bead or a metal nut is suspended by a thin piece of string or thread from the right-angle corner of the quadrant by means of a drawing-pin. When one edge of the quadrant is held horizontally the thread will hang vertically along the other straight edge like a plumb-line. Mark this point on the curved edge o degrees. You are then holding the quadrant horizontally and its angle of tilt is therefore o degrees. To find your latitude you point the quadrant to the Pole Star so that you can just see it along the straight edge. The weighted line will still hang vertically but it will now show on the curved edge of the quadrant what your latitude is.

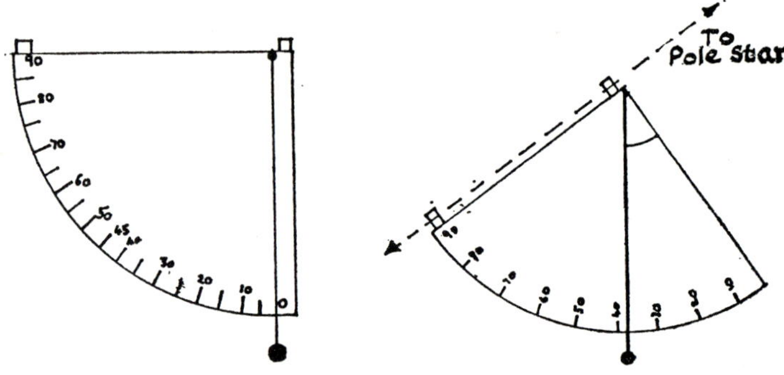

FIG 24 *The Quadrant*

The explanation of the quadrant is quite simple. On the earth's equator the Pole Star is on the horizon, to the north. You would then be holding the quadrant with the straight sighting edge in a horizontal position and the reading would be o degrees. At the north pole the Pole Star would be overhead and the straight edge would be pointing vertically to the zenith, the overhead point. The reading would then be 90 degrees. Most of us live somewhere between these two extremes of latitude and, in the British Isles, the reading will be between fifty and sixty degrees north. Try finding your latitude in this way and check your reading from the latitude shown for your location by a map. Long ago sailors used to find their latitude at sea in this way or by finding the altitude of the sun at noon. Using the sun, however, required some additional calculations because, unlike the Pole Star, its noon altitude varies throughout the year.

Bibliography

General Works of Reference on Astronomy
Concise Encyclopedia of Astronomy *by P. Muller* (Collins)
A Field Guide to the Stars and Planets *by D. Menzel* (Collins)
Stars *by Widmann and Schutte* (Thames & Hudson)
Observer's Book of Astronomy *by P. Moore* (Warne)
Astronomy *by P. Moore* (Oldbourne)
Astronomy *by F. Hoyle* (Macdonald)
1001 Questions Answered about Astronomy *by J. Pickering* (Lutterworth
 Press)
The Amateur Astronomer *by P. Moore* (Lutterworth Press)
Astronomy for Amateurs *by J. Muirden* (Cassell)
The Young Astronomer's Companion *by K. Fea* (Souvenir Press)
Starcraft *by W. Barton and J. Joseph* (Whittlesey House)
The Stars *by I. Adler* (Dobson)
What Star is That? *by P. L. Brown* (Thames & Hudson)

Telescopes and Instruments
Making and Using a Telescope *by P. Moore & H. Wilkins* (Eyre &
 Spottiswoode)
Book of the Telescope *by A. Frank* (Charles Frank)
The Amateur Astronomer & His Telescope *bv G. Roth* (Faber)
Handbook of Telescope-making *by N. Howard* (Faber)
Constructing an Astronomical Telescope *by N. Mathewson* (Blackie)
Making Your Own Telescope *by A. Thompson* (Sky Publishing
 Corporation)
Amateur Astronomer's Handbook *by J. B. Sidgwick* (Faber)
History of the Telescope *by H. C. King* (Griffin)
Amateur Telescope-Making (Vols I, II and III) edited *by Ingalls*
 (Scientific American)
Practical Amateur Astronomy edited *by P. Moore* (Lutterworth Press)

Observation and Telescopic Work
Astronomy With Binoculars *by J. Muirden* (Faber)
Half-hours With the Telescope *by R. A. Proctor* (Longmans)
Celestial Objects for Common Telescopes *by T. W. Webb* (Dover)
A Field Book of the Stars *by W. Olcott* (Putnam)
In Starland With a Three-inch Telescope *by W. Olcott* (Putnam)
Popular Telescopic Astronomy *by A. Fowler* (Philip)
Hours With a Three-inch Telescope *by W. Noble* (Longmans)
Through My Telescope *by W. Hay* (Murray)
How To Enjoy The Starry Sky *by M. Woodward* (Hodder & Stoughton)
1001 Celestial Wonders *by C. Barnes* (Pacific Science Press)
Astronomy for Amateurs ed. *by J. Oliver* (Longmans)
Telescopic Work for Starlight Evenings *by W. Denning* (Taylor &
    Francis)
Astronomy With an Opera-glass *by G. Serviss* (Appleton)
Pleasures of the Telescope *by G. Serviss* (Appleton)
Round the Year With the Stars *by G. Serviss* (Harper)
Astronomy With the Naked Eye *by G. Serviss* (Harper)
Night Skies of the Year *by R. Worvill* (Kahn & Averill)

Some of the above titles are out of print but can be found in libraries
and secondhand bookshops. These are out of date in some particulars
but still contain much of value and interest to the beginner.

Star Atlases and Moon Maps
Norton's Star Atlas (Gall & Inglis)
The same publisher also issues a simplified version under the title of
New Popular Star Atlas.
Chart of the Stars *by E. O. Tancock* (Philip)
Night Sky Map (Daily Telegraph)
Times Book of the Night Sky (Times Newspapers Ltd). An invaluable
little book published annually with star-maps and current information.
Planispheres are made in pocket and larger sizes by the publishing firm
of G. Philip.
Elger's Map of the Moon revised *by H. P. Wilkins* (Philip)
Amateur Astronomer's Photographic Lunar Atlass *by Hatfield* (Lutter-
    worth Press)
Atlas of the Moon *by V. de Callatay* (Macmillan)

Secondhand and new telescopes and equipment are advertised in the columns of the weekly paper, Exchange & Mart, and also in the Journal of the British Astronomical Association (Burlington House, Piccadilly, London). The Association has telescopes and other instruments which are available on loan to members. The official in charge of these is the Association's Curator of Instruments. There is also a section of the Association which is concerned with advising members about telescopes, and other astronomical equipment. The official in charge of this department is the Director of Instruments and Observing Methods.

The Junior Astronomical Society is an associated body which caters for the beginner, regardless of age. Most large towns have local societies which always welcome new members. Addresses of these can be obtained from local public libraries or other offices of information.

The monthly periodical, *Sky and Telescope*, is a valuable source of information for amateurs and is obtainable in most countries. Many makers and suppliers of equipment advertise in its pages although these are intended mainly for readers in the United States of America. In most countries, however, there are astronomical groups or societies, many of which are affiliated to the British Astronomical Association which also has individual members in all parts of the world. Any beginner who wishes to make contact with an affiliated group or a member in his locality should apply to the Assistant Secretary of the Association at the address given above.